Transition to Higher Mathematics

Structure and Proof

Bob A. Dumas
John E. McCarthy

 **Higher Education**

Boston   Burr Ridge, IL   Dubuque, IA   New York   San Francisco   St. Louis
Bangkok   Bogotá   Caracas   Kuala Lumpur   Lisbon   London   Madrid   Mexico City
Milan   Montreal   New Delhi   Santiago   Seoul   Singapore   Sydney   Taipei   Toronto

Higher Education

TRANSITION TO HIGHER MATHEMATICS: STRUCTURE AND PROOF

1 2 3 4 5 6 7 8 9 0 DOC/DOC 0 9 8 7 6

ISBN-13 978–0–07–353353–7
ISBN-10 0–07–353353–X

Publisher: *Elizabeth J. Haefele*
Senior Sponsoring Editor: *Elizabeth Covello*
Developmental Editor: *Dan Seibert*
Executive Marketing Manager: *Kelly R. Brown*
Project Manager: *April R. Southwood*
Senior Production Supervisor: *Sherry L. Kane*
Designer: *John A. Joran*
Cover Illustration: *"Diagonal enumeration of enumerable sets."*
Compositor: *Lachina Publishing Services*
Typeface: *11/13 Times New Roman*
Printer: *R. R. Donnelley Crawfordsville, IN*

Library of Congress Cataloging-in-Publication Data

Dumas, Bob A.
    Transition to higher mathematics : structure and proof / Bob A. Dumas, John E. McCarthy. — 1st ed.
        p.   cm. — (Walter Rudin student series in advanced mathematics)
    Includes index.
    ISBN 978–0–07–353353–7 — ISBN 0–07–353353–X (acid-free paper)
    1. Logic, Symbolic and mathematical—Textbooks. I. McCarthy, John E. (John Edward), 1964–. II. Title.
III. Series.

    QA9.D863 2007
    511.3—dc22   2006043848
                                                                       CIP
www.mhhe.com

To Gloria, Siena, and William
B. D.

To Suzanne, Fiona, and Myles
J. M^{c}C.

Titles in the Walter Rudin Student Series in Advanced Mathematics

Walter Rudin Student Series in Advanced Mathematics
Editorial Board

Contents

CHAPTER 0

Introduction

0.1 Why This Book Is

More students today than ever before take calculus in high school. This comes at a cost, however: fewer and fewer take a rigorous course in Euclidean geometry. Moreover, the calculus course taken by almost all students, whether in high school or college, avoids proofs and often does not even give a formal definition of a limit. Indeed some students enter the university having never read or written a proof by induction or encountered a mathematical proof of any kind.

As a consequence, teachers of upper-level undergraduate mathematics courses in linear algebra, abstract algebra, analysis, and topology have to work extremely hard inculcating the concept of proof while simultaneously trying to cover the syllabus. This problem has been addressed at many colleges and universities by introducing a bridge course, with a title like "Foundations for Higher Mathematics," taken by students who have completed the regular calculus sequence. Some of these students plan to become mathematics majors. Others just want to learn some more mathematics; but if what they are exposed to is interesting and satisfying, many will choose to major or double major in mathematics.

This book is written for students who have taken calculus and want to learn what "real mathematics" is. We hope you will find the material engaging and interesting, and that you will be encouraged to learn more advanced mathematics.

0.2 What This Book Is

The purpose of this book is to introduce you to the culture, language, and thinking of mathematicians. We say "mathematicians," not "mathematics," to emphasize that mathematics is, at heart, a human endeavor. If there is intelligent life in Erewhemos, then the Erewhemosians will surely agree that two plus two equals four. If they have thought carefully about the question, they will not believe that the square root of two can be exactly given by the ratio of two whole numbers or that there are finitely many prime numbers. However, we can only speculate about whether they would find these latter questions remotely interesting or what they might consider satisfying answers to questions of this kind.

Mathematicians have, after millennia of struggles and arguments, reached a widespread (if not quite universal) agreement as to what constitutes an acceptable mathematical argument. They call this a "proof," and it constitutes a carefully reasoned argument based on agreed premises. The methodology of mathematics has been spectacularly successful, and it has spawned many other fields. In the twentieth century, computer programming and applied statistics developed from offshoots of mathematics into disciplines of their own. In the nineteenth century, so did astronomy and physics. The increasing availability of data make the treatment of data in a sophisticated mathematical way one of the great scientific challenges of the twenty-first century.

In this book, we shall try to teach you what a proof is—what level of argument is considered convincing, what is considered over-reaching, and what level of detail is considered too much. We shall try to teach you how mathematicians think—what structures they use to organize their thoughts. A structure is like a skeleton—if you strip away the inessential details you can focus on the real problem. A great example of this is the idea of number, the earliest human mathematical structure. If you learn how to count apples and that two apples plus two apples make four apples, and if you think that this is about apples rather than counting, then you still do not know what two sheep plus two sheep make. But once you realize that there is an underlying structure of number and that two plus two is four in the abstract, then adding wool or legs to the objects does not change the arithmetic.

0.3 What This Book Is Not

There is an approach to teaching a transition course that many instructors favor. It is to have a problem-solving course in which students learn to write proofs in a context where their intuition can help, such as in combinatorics or number theory. This helps to make the course interesting and can keep students from getting totally lost.

We have not adopted this approach. Our reason is that in addition to teaching the skill of writing a logical proof, we also want to teach the skill of carefully analyzing definitions. Much of the instructor's labor in an upper-division algebra or analysis course consists of forcing the students to carefully read the definitions of new and unfamiliar objects, to decide which mathematical objects satisfy the definition and which do not, and to understand what follows "immediately" from the definitions. Indeed, the major reason that the epsilon-delta definition of limit has disappeared from most introductory calculus courses is the difficulty of explaining how the quantifiers $\forall \varepsilon \; \exists \delta$, in precisely this order, give the exact notion of limit for which we are striving. Thus, while students must work harder in this course to learn more abstract mathematics, they will be better prepared for advanced courses.

Nor is this a text in applied logic. The early chapters of the book introduce the student to the basic mathematical structures through formal definitions. Although we provide a rather formal treatment of first-order logic and mathematical induction, our objective is to move to more advanced classical *mathematical* structures and arguments as soon as the student has an adequate understanding of the logic underlying mathematical proofs.

0.4 Advice to the Student

Welcome to higher mathematics! If your exposure to college-level mathematics is limited to calculus, this book will probably seem very different from your previous texts. Many students learn calculus by quickly scanning the text and proceeding directly to the problems. When struggling with a problem, they seek similar problems in the text and attempt to emulate the solution they find. Finally, they check the solution, usually found at the back of the text, to "validate" the methodology.

This book, like many texts addressing more advanced topics, is not written with computational problems in mind. Our objective is to introduce you to the various elements of higher undergraduate mathematics— the culture, language, methods, topics, standards, and results. The problems in these courses are to prove true mathematical claims or refute untrue claims. In the context of calculus, the mathematician must prove the results that you freely used. To most people, this activity seems very different from computation. For instance, you will probably find it necessary to think about a problem for some time before you begin writing. Unlike calculus, in which the general direction of the methods is usually obvious, trying to prove mathematical claims can feel directionless or accidental. However, it is strategic rather than random. This is one of the great challenges of mathematics—at the higher levels, it is creative, not rote. With practice and disciplined thinking, you will learn to see your way to proving mathematical claims.

We shall begin our treatment of higher mathematics with a large number of definitions. This is usual in a mathematics course and is necessary because mathematics requires precise expression. We shall try to motivate these definitions so that their usefulness will be obvious as early as possible. After presenting and discussing some definitions, we shall present arguments for some elementary claims concerning these definitions. This will give us some practice in reading, writing, and discussing mathematics. In the early chapters of the book, we include numerous discussions and remarks to help you grasp the basic direction of the arguments. In the later chapters of the book, you will read more difficult arguments for some deep classical results. We recommend that you read these arguments deliberately to ensure your thorough understanding of the argument and to nurture your sense of the level of detail and rigor expected in an undergraduate mathematical proof.

There are exercises at the end of each chapter designed to direct your attention to the reading and compel you to think through the details of the proofs. Some of these exercises are straightforward, but many of them are very hard. We do not expect that every student will be able to solve every problem. However, spending an hour (or more) thinking about a difficult problem is time well spent even if you do not solve the problem: it strengthens your mathematical muscles and allows you to appreciate, and to understand more deeply, the solution if it is eventually shown to you. Ultimately, you will be able to solve some of the hard problems yourself after thinking deeply about them. Then you will be a real mathematician!

Mathematics is, from one point of view, a logical exercise. We define objects that do not physically exist and use logic to draw the deepest conclusions we can concerning these objects. If this were the end of the story, mathematics would be no more than a game and would be of little enduring interest. It happens, however, that interpreting physical objects, processes, behaviors, and other subjects of intellectual interest as mathematical objects, and applying the conclusions and techniques from the study of these mathematical objects, allows us to draw reliable and powerful conclusions about practical problems. This method of using mathematics to understand the world is called mathematical modelling. The world in which you live, the way you understand this world, and how it differs from the world and understanding of your distant ancestors is to a large extent the result of mathematical investigation. In this book, we try to explain how to draw mathematical conclusions with certainty. When you studied calculus, you used numerous deep theorems in order to draw conclusions that otherwise might have taken months rather than minutes. Now we shall develop an understanding of how results of this depth and power are derived.

0.5 Advice to the Instructor

Learning terminology—what do "contrapositive" and "converse" mean—comes easily to most students. Your challenge in the course is to teach them how to read definitions closely and then how to manipulate them. This is much harder when there is no concrete image that students can keep in mind. Vectors in $\mathbb{R}^n$, for example, are more intimidating than in $\mathbb{R}^3$, not because of any great inherent increase in complexity but because they are harder to think of geometrically, so students must trust the algebra alone. This trust takes time to build.

Chapter 1 is mainly to establish notation and discuss necessary concepts that some may have already seen (like injections and surjections). Unfortunately this may be the first exposure to some of these ideas for many students, so the treatment is rather lengthy. The speed at which the material is covered naturally will depend on the strength and background of the students. Take some time explaining why a sequence can be thought of as a function with domain $\mathbb{N}$—variations on this idea will recur.

Chapter 2 introduces relations. These are hard to grasp because of the abstract nature of the definition. Equivalences and linear orderings recur throughout the book, and students' comfort with these will increase.

Neither Chapter 1 nor Chapter 2 dwell on proofs. In fact mathematical proofs and elementary first-order logic are not introduced until Chapter 3. Our objective is to get students thinking about mathematical structures and definitions without the additional psychic weight of reading and writing proofs. We use examples to illustrate the definitions. The first chapters provide basic conceptual foundations for later chapters, and we find that most students have their hands full just trying to read and understand the definitions and examples. In the exercises we ask the students to "show" the truth of some mathematical claims. Our intention is to get the student thinking about the task of proving mathematical claims. It is not expected that they will write successful arguments before Chapter 3. We encourage the students to attempt the problems even though they will likely be uncertain about the requirements for a mathematical proof. If you feel strongly that mathematical proofs need to be discussed before launching into mathematical definitions, you can cover Chapter 3 first.

Chapter 3 is fairly formal, and should go quickly. Chapter 4 introduces students to the first major proof technique—induction. With practice, they can be expected to master this technique. We also introduce as an ongoing theme the study of polynomials and prove, for example, that a polynomial has no more roots than its degree.

Chapters 5, 6, and 7 are completely independent of each other. Chapter 5 treats limits and continuity, up to proving that the uniform limit of a sequence of continuous functions is continuous. Chapter 6 is on infinite sets, proving Cantor's theorems and the Schröder-Bernstein theorem. By the end of the chapter, the students will have come to appreciate that it is generally much easier to construct two injections than one bijection!

Chapter 7 contains a little number theory—up to the proof of Fermat's little theorem. It then shows how much of the structure transfers to the algebra of real polynomials.

Chapter 8 constructs the real numbers, using Dedekind cuts, and proves that they have the least upper bound property. This is then used to prove the basic theorems of real analysis—the Intermediate Value theorem and the Extreme Value theorem. Sections 8.1–8.4 require only Chapters 1–4 and Section 6.1. Sections 8.5–8.8 require Sections 5.1 and 5.2. Section 8.9 requires Chapter 6.

In Chapter 9, we introduce the complex numbers. Sections 9.1–9.3 prove the Tartaglia-Cardano formula for finding the roots of a cubic and point out how it is necessary to use complex numbers even to find real roots of real cubics. These sections require only Chapters 1–4. In Section 9.4 we prove the Fundamental Theorem of Algebra. This requires Chapter 5 and the Bolzano-Weierstrass theorem from Section 8.6.

What is a reasonable course based on this book? Chapters 1–4 are essential for any course. In a one-quarter course, one could also cover Chapter 6 and either Chapter 5 or 7. In a semester-long course, one could cover Chapters 1–6 and one of the remaining three chapters. Chapter 9 can be covered without Chapter 8 if one is willing to assert the least upper bound property as an axiom of the real numbers, and then Section 8.6 can be covered before Section 9.4 without any other material from Chapter 8.

We suggest that you agree with your colleagues on a common curriculum for this course so that topics that you cover thoroughly (e.g., cardinality) need not be repeated in successive courses.

This transition course is becoming one of the most important courses in the mathematics curriculum and the first important course for the mathematics major. For the talented and intellectually discriminating first- or second-year student the standard early courses in the mathematics curriculum—calculus, differential equations, matrix algebra—provide little incentive for studying mathematics. Indeed, there is little mathematics in these courses, and less still with the evolution of lower undergraduate curricula toward the service of the sciences and engineering. This is particularly disturbing as it pertains to the talented student who has not yet decided on a major and may never have considered mathematics. We believe that the best students should be encouraged to take this course as early as possible—even concurrent with the second semester or third quarter of first-year calculus. It is not just to help future math majors but can also serve a valuable rôle in recruiting them by letting smart students see that mathematics is challenging and, more to the point, interesting and deep. Mathematics is its own best apologist. Expose the students early to authentic mathematical thinking and results, and let them make an informed choice. It may come as a surprise to some, but good students still seek what mathematicians sought as students—the satisfaction of mastering a difficult, interesting, and useful discipline.

0.6 Acknowledgments

We have received a lot of help in writing this book. In addition to the support of our families, we have received valuable advice and feedback from our students and colleagues and from the reviewers of the manuscript. In particular we would like to thank Matthew Valeriote for many helpful discussions and Alexander Mendez for drawing all the figures in the book.

Additional thanks go to the following manuscript reviewers:

- Lowell Abrams, George Washington University
- Gerald Beer, California State University–Los Angeles
- H. Lamar Bentley, University of Toledo
- Michael Berg, Loyola Marymount University
- James Campbell, University of Memphis
- George Davis, Georgia State University
- Daniel Dreibelbis, University of North Florida
- John Drew, College of William and Mary
- Aniekan Ebiefung, University of Tennessee–Chattanooga
- Michael Falk, Northern Arizona University
- Maria Girardi, University of South Carolina
- John Koker, University of Wisconsin–Oshkosh
- Edward Letzter, Temple University
- Marvin Mielke, University of Miami
- Efton Park, Texas Christian University
- Mihai Putinar, University of California–Santa Barbara
- Laura Schoppmann, Seton Hall University
- Joe Sharp, State University of West Georgia
- Roy Smith, University of Georgia
- Katherine Stevenson, University of California–Northridge
- Mark Watkins, Syracuse University
- Paul Zorn, St. Olaf College

CHAPTER 1

Preliminaries

To communicate mathematics you will need to understand and abide by the conventions of mathematicians. In this chapter we review some of these conventions.

1.1 "And" and "Or"

Statements are declarative sentences; that is, a statement is a sentence that is true or false. Mathematicians make mathematical statements—sentences about mathematics that are true or false. For instance, the statement

> All prime numbers, except the number 2, are odd

is a true statement. The statement

$$3 < 2$$

is false.

We use natural language connectives to combine mathematical statements. The connectives "and" and "or" have a particular usage in mathematical prose. Let P and Q be mathematical statements. The statement

> P and Q

is the statement that both P and Q are true.

Mathematicians use what is called the "inclusive or." In everyday usage the statement "P or Q" can sometimes mean that exactly one (but not both) of the statements P and Q is true. In mathematics, the statement

$$P \text{ or } Q$$

is true when either or both statements are true, that is, when any of the following hold:

P is true and Q is false.

P is false and Q is true.

P is true and Q is true.

1.2 Sets

Intuitively, a mathematical set is a collection of mathematical objects. Unfortunately this simple characterization of sets, carelessly handled, gives rise to contradictions. Some collections will turn out not to have the properties that we demand of mathematical sets. An example of how this can occur is presented in Section 1.7. We shall not develop formal set theory from scratch here. Instead, we shall assume that certain building block sets are known, and we shall describe ways to build new sets out of these building blocks.

Our initial building blocks will be the sets of natural numbers, integers, rational numbers, and real numbers. In Chapter 8, we shall show how to build all these from the natural numbers. One cannot go much further than this, though: in order to do mathematics, one has to start with axioms that assert that the set of natural numbers exists.

DEFINITION. Element, $\in$ If X is a set and x is an object in X, we say that x is an element, or *member*, of X. This is written

$$x \in X.$$

We write $x \notin X$ if x is not a member of X.

There are numerous ways to define sets. If a set has few elements, it may be defined by listing. For instance,

$$X = \{2, 3, 5, 7\}$$

is the set of the first four prime numbers. In the absence of any other indication, a set defined by a list is assumed to have as elements only the objects in the list. For sets with too many elements to list, we must provide the reader with a means to determine membership in the set. The author can inform the reader that not all elements of the set have been listed but that enough information has been provided for the reader to identify a pattern for determining membership in the set. For example, let

$$X = \{2, 4, 6, 8, \ldots, 96, 98\}.$$

Then X is the set of positive even integers less than 100. However, using an ellipsis to define a set may not always work: it assumes that the reader will identify the pattern you wish to characterize. Although this usually works, it carries the risk that the reader is unable to correctly identify the pattern intended by the author.

Some sets are so important that they have standard names and notations that you will need to know.

NOTATION. Natural numbers, $\mathbb{N}$ The natural numbers are the elements of the set

$$\{0, 1, 2, 3, \ldots\}.$$

This set is denoted by $\mathbb{N}$.

Warning: Many authors call $\{1, 2, 3, \ldots\}$ the set of natural numbers. This is a matter of definition, and there is no universal convention; logicians tend to favor our convention and algebraists the other. In this book, we shall use $\mathbb{N}^+$ to denote $\{1, 2, 3, \ldots\}$.

NOTATION. $\mathbb{N}^+$ $\mathbb{N}^+$ is the set of positive integers

$$\{1, 2, 3, \ldots\}.$$

NOTATION. Integers, $\mathbb{Z}$ $\mathbb{Z}$ is the set of integers

$$\{\ldots, -3, -2, -1, 0, 1, 2, 3, \ldots\}.$$

NOTATION. Rational numbers, $\mathbb{Q}$　$\mathbb{Q}$ is the set of rational numbers

$$\left\{ \frac{p}{q} \text{ where } p, q \in \mathbb{Z} \text{ and } q \neq 0 \right\}.$$

NOTATION. Real numbers, $\mathbb{R}$　$\mathbb{R}$ is the set of real numbers.

A good understanding of the real numbers requires a bit of mathematical development. In fact, it was only in the nineteenth century that we really came to a modern understanding of $\mathbb{R}$. We shall have a good deal to say about real numbers in Chapter 8.

DEFINITION. A number x is *positive* if $x > 0$. A number x is *nonnegative* if $x \geq 0$.

NOTATION. X^+　If X is a set of real numbers, we use X^+ for the positive numbers in the set X.

The notation we have presented for these sets is widely used. We introduce a final convention for set names that is not as widely recognized but is useful for set theory.

NOTATION. $\ulcorner n \urcorner$ is the set of all natural numbers less than n:

$$\ulcorner n \urcorner = \{0, 1, 2, \ldots, n - 1\}.$$

One purpose of this notation is to canonically associate any natural number n with a set having exactly n elements.

The reader should note that we have not *defined* the above sets. We are assuming that you are familiar with them, and some of their properties, by virtue of your previous experience in mathematics. We shall eventually define the sets systematically in Chapter 8.

A more precise method of defining a set is to use unambiguous conditions that characterize membership in the set.

NOTATION. $\{x \in X \mid P(x)\}$ Let X be a (previously defined) set, and let $P(x)$ be a condition or property. Then the set

$$Y = \{x \in X \mid P(x)\} \qquad\qquad (1.1)$$

is the set of elements in X that satisfy condition P. The set X is called the domain of the variable.

In words, (1.1) is read: "Y equals the set of all (lowercase) x in (uppercase) X such that P is true of x." The symbol " $\mid$ " in (1.1) is often written instead with a colon, namely $\{x \in X : P(x)\}$. In mathematics, $P(x)$ is a often a mathematical formula. For instance, suppose $P(x)$ is the formula "$x^2 = 4$." By $P(2)$ we mean the formula with 2 substituted for x, that is

$$2^2 = 4.$$

If the substitution results in a true statement, we say that $P(x)$ holds at 2, or $P(2)$ is true. If the statement that results from the substitution is false, for instance $P(1)$, we say that $P(x)$ does not hold at 1, or that $P(1)$ is false.

EXAMPLE 1.2. Consider the set

$$X = \{0, 1, 4, 9, \ldots\}.$$

A precise definition of the same set is the following:

$$X = \{x \in \mathbb{N} \mid \text{for some } y \in \mathbb{N}, \ x = y^2\}.$$

EXAMPLE 1.3. Let Y be the set of positive even integers less than 100. Then Y can be written:

$$\{x \in \mathbb{N} \mid x < 100 \text{ and there is } n \in \mathbb{N}^+ \text{ such that } x = 2 \cdot n\}.$$

EXAMPLE 1.4. An interval I is a nonempty subset of $\mathbb{R}$ with the property that whenever $a, b \in I$ and $a < c < b$, then c is in I. A bounded interval must have one of the four forms:

$$
\begin{aligned}
(a,b) &= \{x \in \mathbb{R} \mid a < x < b\} \\
[a,b) &= \{x \in \mathbb{R} \mid a \le x < b\} \\
(a,b] &= \{x \in \mathbb{R} \mid a < x \le b\} \\
[a,b] &= \{x \in \mathbb{R} \mid a \le x \le b\}
\end{aligned}
$$

where, in the first three cases, a and b are real numbers with $a < b$; in the fourth case, we just require $a \le b$. Unbounded intervals have five forms:

$$
\begin{aligned}
(-\infty, b) &= \{x \in \mathbb{R} \mid x < b\} \\
(-\infty, b] &= \{x \in \mathbb{R} \mid x \le b\} \\
(b, \infty) &= \{x \in \mathbb{R} \mid x > b\} \\
[b, \infty) &= \{x \in \mathbb{R} \mid x \ge b\} \\
\mathbb{R} &
\end{aligned}
$$

where b is some real number. An interval is called *closed* if it contains all its endpoints (both a and b in the first group of examples, just b in the first four examples of the second group) and *open* if it contains none of them. Notice that this makes $\mathbb{R}$ the only interval that is both closed and open.

For the sake of brevity, an author may not explicitly identify the domain of the variable. Be careful about this, as the author is relying on the reader to make the necessary assumptions. For instance, consider the set

$$X = \{x \mid (x^2 - 2)(x - 1)(x^2 + 1) = 0\}.$$

If the domain of the variable is assumed to be $\mathbb{N}$, then

$$X = \{1\}.$$

If the domain of the variable is assumed to be $\mathbb{R}$, then

$$X = \{1, \sqrt{2}, -\sqrt{2}\}.$$

If the domain of the variable is assumed to be the complex numbers, then

$$X = \{1, \sqrt{2}, -\sqrt{2}, i, -i\}$$

where i is the complex number $\sqrt{-1}$. Remember, the burden of clear communication is on the author, not the reader.

Another alternative is to include the domain of the variable in the condition that defines membership in the set. So, if X is the intended domain of the set and $P(x)$ is the condition for membership in the set, then

$$\{x \in X \mid P(x)\} = \{x \mid x \in X \text{ and } P(x)\}.$$

As long as the definition is clear, the author has some flexibility with regard to notation.

1.2.1 Set Identity. When are two sets equal? You might be inclined to say that two sets are equal provided they are the *same* collection of objects. Of course this is true, but equality as a relation between objects is not very interesting. However, you have probably spent a lot of time investigating equations (which are just statements of equality), and we doubt that equality seemed trivial. This is because, in general, equality should be understood as a relationship between *descriptions* or *names* of objects rather than between the objects themselves. The statement

$$a = b$$

is a claim that the object represented by a is the same object as that represented by b. For example, the statement

$$5 - 3 = 2$$

is the claim that the number represented by the arithmetic expression $5 - 3$ is the same number as that represented by the numeral 2.

In the case of sets, this notion of equality is called *extensionality*.

DEFINITION. Extensionality Let X and Y be sets. Then $X = Y$ provided that every element of X is also an element of Y and every element of Y is also an element of X.

There is flexibility in how a set is characterized as long as we are clear on which objects constitute the set. For instance, consider the set equation

$$\{\,\text{Mark Twain, Samuel Clemens}\} = \{\text{Mark Twain}\}.$$

If by "Mark Twain" and "Samuel Clemens" we mean the deceased American author, these sets are equal, by extensionality, and the statement is true. The set on the left-hand side of the equation has only one element since both names refer to the same person. If, however, we consider "Mark Twain" and "Samuel Clemens" as names, the statement is false since "Samuel Clemens" is a member of the set on the left-hand side of the equation but not the right-hand side. You can see that set definitions can depend on the implicit domain of the variable even if the sets are defined by listing.

EXAMPLE 1.5. Consider the following six sets:

$$
\begin{aligned}
X_1 &= \{1, 2\} \\
X_2 &= \{2, 1\} \\
X_3 &= \{1, 2, 1\} \\
X_4 &= \{n \in \mathbb{N} \mid 0 < n < 3\} \\
X_5 &= \{n \in \mathbb{N} \mid \text{there exist } x, y, z \in \mathbb{N}^+ \text{ such that } x^n + y^n = z^n\} \\
X_6 &= \{0, 1, 2\}.
\end{aligned}
$$

The first five sets are all equal, and the sixth is different. However, while it is obvious that $X_1 = X_2 = X_3 = X_4$, the fact that $X_5 = X_1$ is the celebrated theorem of Andrew Wiles (his proof of Fermat's last theorem).

1.2.2 Relating Sets. In order to say anything interesting about sets, we need ways to relate them, and we shall want ways to create new sets from existing sets.

DEFINITION. Subset, $\subseteq$ Let X and Y be sets. X is a subset of Y if every element of X is also an element of Y. This is written

$$X \subseteq Y.$$

DEFINITION. Superset, $\supseteq$ If $X \subseteq Y$, then Y is called a superset of X, written

$$Y \supseteq X.$$

To show that two sets are equal (or that two descriptions of sets refer to the same set), you must show that they have precisely the same elements. It is often easier if the argument is broken into two simpler arguments in which you show mutual containment of the sets. In other words, saying $X = Y$ is the same as saying

$$X \subseteq Y \text{ and } Y \subseteq X \tag{1.6}$$

and verifying the two separate claims in (1.6) is often easier (or at least clearer) than showing that $X = Y$ all at once.

Let us add a few more elementary notions to our discussion of sets.

DEFINITION. Proper subset, $\subsetneq$, $\supsetneq$ Let X and Y be sets. X is a proper subset of Y if

$$X \subseteq Y \text{ and } X \neq Y.$$

We write this as

$$X \subsetneq Y$$

or

$$Y \supsetneq X.$$

DEFINITION. Empty set, $\emptyset$ The empty set is the set with no elements. It is denoted by $\emptyset$.

So for any set X,

$$\emptyset \subseteq X.$$

(Think about why this is true.) Just because $\emptyset$ is empty does not mean it is unimportant. Indeed, many mathematical questions reduce to asking whether a particular set is empty or not. Furthermore, as you will see in Chapter 8, we can build the entire real line from the empty set using set operations.

EXERCISE. (See Exercise 1.1). Show that

$$\{n \in \mathbb{N} \mid n \text{ is odd and } n = k(k+1) \text{ for some } k \in \mathbb{N}\}$$

is empty.

Let us discuss some ways to define new sets from existing sets.

DEFINITION. Union, $\cup$ Let X and Y be sets. The union of X and Y, written $X \cup Y$, is the set

$$X \cup Y = \{x \mid x \in X \text{ or } x \in Y\}.$$

(Recall our discussion in Section 1.1 about the mathematical meaning of the word "or.")

DEFINITION. Intersection, $\cap$ Let X and Y be sets. The intersection of X and Y, written $X \cap Y$, is the set

$$X \cap Y = \{x \mid x \in X \text{ and } x \in Y\}.$$

DEFINITION. Set difference, \ Let X and Y be sets. The set difference of X and Y, written $X \setminus Y$, is the set

$$X \setminus Y = \{x \in X \mid x \notin Y\}.$$

DEFINITION. Disjoint Let X and Y be sets. X and Y are disjoint if

$$X \cap Y = \emptyset.$$

Oftentimes one deals with sets that are subsets of some fixed given set U. For example, when dealing with sets of natural numbers, the set U would be $\mathbb{N}$.

DEFINITION. Complement Let $X \subseteq U$. The complement of X in U is the set $U \setminus X$. When U is understood from the context, the complement of X is written X^c.

What about set operations involving more than two sets? Unlike arithmetic, in which there is a default order of operations (powers, products, and sums), there is not a universal convention for the order in which set operations are performed. If intersections and unions appear in the same expression, then the order in which the operations are performed can matter. For instance, suppose X and Y are disjoint, nonempty sets, and consider the expression

$$X \cap X \cup Y.$$

If we mean for the intersection to be executed before the union, then

$$(X \cap X) \cup Y = X \cup Y.$$

If, however, we intend the union to be computed before the intersection, then

$$X \cap (X \cup Y) = X.$$

Since Y is nonempty and disjoint from X,

$$(X \cap X) \cup Y \neq X \cap (X \cup Y).$$

Consequently, the order in which set operations are executed needs to be explicitly prescribed with parentheses.

EXAMPLE 1.7. Let $X = \mathbb{N}$ and $Y = \mathbb{Z} \setminus \mathbb{N}$. Then

$$(X \cap X) \cup Y = \mathbb{N} \cup Y = \mathbb{Z}.$$

However

$$X \cap (X \cup Y) = \mathbb{N} \cap \mathbb{Z} = \mathbb{N}.$$

DEFINITION. Cartesian product, Direct product, $X \times Y$ Let X and Y be sets. The Cartesian product of X and Y, written $X \times Y$, is the set of ordered pairs

$$\{(x, y) \mid x \in X \text{ and } y \in Y\}.$$

The Cartesian product is also called the direct product.

EXAMPLE 1.8. Let

$$X = \{1, 2, 3\}$$

and

$$Y = \{1, 2\}.$$

Then

$$X \times Y = \{(1, 1), (1, 2), (2, 1), (2, 2), (3, 1), (3, 2)\}.$$

Note that the order matters—that is

$$(1, 2) \neq (2, 1).$$

So $X \times Y$ is a set with six elements.

Since direct products are themselves sets, we can easily define the direct product of more than two factors. For example, let X, Y, and Z be sets, then

$$(X \times Y) \times Z = \{((x, y), z) \mid x \in X, y \in Y, z \in Z\}.$$

Formally,

$$(X \times Y) \times Z \neq X \times (Y \times Z) \qquad (1.7)$$

because $((x, y), z)$ and $(x, (y, z))$ are not the same. However in nearly every application, this distinction is not important, and mathematicians generally consider the direct product of more than two sets without regard to this detail. Therefore you will generally see the Cartesian product of three sets written without parentheses,

$$X \times Y \times Z.$$

In this event you may interpret the direct product as either side of statement 1.7.

With some thought, you can conclude that we have essentially described the Cartesian product of an arbitrary finite collection of sets. The elements of the Cartesian product $X \times Y$ are ordered pairs. Our characterization of the Cartesian product of three sets, X, Y, and Z, indicates that its elements could be thought of as ordered pair of elements of $X \times Y$ and Z, respectively. From a practical point of view, it is simpler to think of elements of $X \times Y \times Z$ as ordered triples. We generalize this as follows.

DEFINITION. Cartesian product, Direct product, $\prod_{i=1}^{n} X_i$ Let $n \in \mathbb{N}^+$, and let $X_1, X_2, \ldots, X_n$ be sets. The Cartesian product of $X_1, \ldots, X_n$, written $X_1 \times X_2 \times \ldots \times X_n$, is the set

$$\{(x_1, x_2, \ldots, x_n) \mid x_i \in X_i,\ 1 \leq i \leq n\}.$$

This may also be written

$$\prod_{i=1}^{n} X_i.$$

When we take the Cartesian product of a set X with itself n times, we write it as X^n:

$$X^n := \overbrace{X \times X \times \cdots \times X}^{n \text{ times}}.$$

1.3 Functions

Like sets, functions are ubiquitous in mathematics.

DEFINITION. Function, $f : X \to Y$ Let X and Y be sets. A function f from X to Y, denoted by $f : X \to Y$, is an assignment of exactly one element of Y to each element of X.

For each element $x \in X$, the function f associates or selects a unique element $y \in Y$. The uniqueness condition does not allow x to be assigned to distinct elements of Y. It does allow different elements of X to be assigned to the same element of Y however. It is important to your understanding of functions that you consider this point carefully. The following examples may help illustrate this.

EXAMPLE 1.9. Let $f : \mathbb{Z} \to \mathbb{R}$ be given by

$$f(x) = x^2.$$

Then f is a function in which the element of $\mathbb{R}$ assigned to the element x of $\mathbb{Z}$ is specified by the expression x^2. For instance f assigns 9 to the integer 3. We express this by writing

$$f(3) = 9.$$

Observe that not every real number is assigned to a number from $\mathbb{Z}$. Furthermore, observe that 4 is assigned to both 2 and -2. Check that f does satisfy the definition of a function.

EXAMPLE 1.10. Let $g : \mathbb{R} \to \mathbb{R}$ be defined by $g(x) = \tan(x)$. Then g is not a function because it is not defined when $x = \pi/2$ (or whenever $x - \pi/2$ is an integer multiple of π). This can be fixed by defining

$$X = \mathbb{R} \setminus \{\pi/2 + k\pi \mid k \in \mathbb{Z}\}.$$

Then $\tan : X \to \mathbb{R}$ is a function from X to $\mathbb{R}$.

EXAMPLE 1.11. Consider two rules, $f, g : \mathbb{R} \to \mathbb{R}$, defined by

$$f(x) = y \qquad \text{if } 3x = 2 - y$$
$$g(x) = y \qquad \text{if } x = y^4.$$

Then f is a function and can be given explicitly as $f(x) = 2 - 3x$. But g does not define a function, because, for example, when $x = 16$, then $g(x)$ could be either 2 or -2.

DEFINITION. Image Let $f : X \to Y$. If $a \in X$, then the element of Y that f assigns to a is denoted by $f(a)$ and is called the image of a under f.

The notation $f : X \to Y$ is a statement that f is a function from X to Y. This statement has as a consequence that for every $a \in X$, $f(a)$ is a specific element of Y. We give an alternative characterization of functions based on Cartesian products.

DEFINITION. Graph of a function Let $f : X \to Y$. The graph of f, graph(f), is

$$\{(x, y) \mid x \in X \text{ and } f(x) = y\}.$$

EXAMPLE 1.12. Let $X \subseteq \mathbb{R}$ and $f : X \to \mathbb{R}$ be defined by $f(x) = -x$. Then the graph of f is

$$\{(x, -x) \mid x \in X\}.$$

EXAMPLE 1.13. The empty function f is the function with empty graph (i.e., the graph of f is the empty set). This means $f : \emptyset \to Y$ for some set Y.

If $f : X \to Y$, then

$$\text{graph}(f) \subseteq X \times Y.$$

Let $Z \subseteq X \times Y$. Then Z is the graph of a function from X to Y if

(i) for any $x \in X$, there is some y in Y such that $(x, y) \in Z$

(ii) if (x, y) is in Z and (x, z) is in Z, then $y = z$.

Suppose X and Y are subsets of $\mathbb{R}$. Then Condition (i) is the condition that every vertical line through a point of X cuts the graph at least once. Condition (ii) is the condition that every vertical line through a point of X cuts the graph at most once.

DEFINITION. Domain, Codomain   Let $f : X \to Y$. The set X is called the domain of f and is written $\mathrm{Dom}(f)$. The set Y is called the codomain of f.

The domain of a function is a necessary component of the definition of a function. The codomain is a bit more subtle. If you think of functions as sets of ordered pairs, that is, if you identified the function with its graph, then every function would have many possible codomains (take any superset of the original codomain). Set theorists think of functions this way, and if functions are considered as sets, extensionality requires that functions with the same graph be identical. However, this convention would make a discussion of surjections clumsy (see page 31), so we shall not adopt it.

When you write

$$f : X \to Y$$

you are explicitly naming the intended codomain, and this makes the codomain a crucial part of the definition of the function. You are indicating to the reader that your definition includes more than just the graph of the function. The definition of a function includes three pieces: the domain, the codomain, and the graph.

EXAMPLE 1.14. Let $f : \mathbb{N} \to \mathbb{N}$ be defined by

$$f(n) = n^2.$$

Let $g : \mathbb{N} \to \mathbb{R}$ be defined by

$$g(x) = x^2.$$

Then graph(f) = graph(g). If $h : \mathbb{R} \to \mathbb{R}$ is defined by

$$h(x) = x^2$$

then graph$(f) \subsetneq$ graph(h), so $f \neq h$ and $g \neq h$. Although graph(f) = graph(g), we consider f and g to be different functions because they have different codomains.

DEFINITION. Range Let $f : X \to Y$. The range of f, Ran(f), is

$$\{y \in Y \mid \text{ for some } x \in X, \, f(x) = y\}.$$

So if $f : X \to Y$, then Ran$(f) \subseteq Y$ and is precisely the set of images under f of elements in X. That is

$$\text{Ran}(f) = \{f(x) \mid x \in X\}.$$

No proper subset of Ran(f) can serve as a codomain for a function that has the same graph as f.

EXAMPLE 1.15. With the same notation as in Example 1.14, we have Ran(f) = Ran(g) = $\{n \in \mathbb{N} \mid n = k^2$ for some $k \in \mathbb{N}\}$. The range of h is $[0, \infty)$.

DEFINITION. Real-valued function, Real function   Let $f : X \to Y$. If Ran$(f) \subseteq \mathbb{R}$, we say that f is a real-valued function. If $X \subseteq \mathbb{R}$ and f is a real-valued function, then we call f a real function.

It is sometimes said that a function is a *rule* that assigns, to each element of a given set, some element from another set. If, by a rule, one means an instruction of some sort, you will see in Chapter 6 that there are "more" functions that cannot be characterized by rules than there are functions that can be. In practice, however, most of the functions we use are defined by rules.

If a function is given by a rule, it is common to write it in the form

$$f : X \to Y$$
$$x \mapsto f(x).$$

The symbol $\mapsto$ is read "is mapped to." For example, the function g in the previous example could be defined by

$$g : \mathbb{N} \;\to\; \mathbb{R}$$
$$n \;\mapsto\; n^2.$$

EXAMPLE 1.16. The function

$$f : \mathbb{R} \;\to\; \mathbb{R}$$
$$x \;\mapsto\; \begin{cases} 0 & x < 0 \\ x + 1 & x \geq 0 \end{cases}$$

is defined by a rule, even though to apply the rule to a given x you must first check where in the domain x lies.

When a real function is defined by a rule and the domain is not explicitly stated, it is taken to be the largest set for which the rule is defined. This is the usual convention in calculus: real functions are defined by mathematical expressions, and it is understood that the implicit domain of a function is the largest subset of $\mathbb{R}$ for which the expression makes sense. The codomain of a real function is assumed to be $\mathbb{R}$ unless explicitly stated otherwise.

EXAMPLE 1.17. Let $f(x) = \sqrt{x}$ be a real function. The domain of the function is assumed to be

$$\{x \in \mathbb{R} \mid x \geq 0\}.$$

DEFINITION. Operation Let X be a set, and $n \in \mathbb{N}^+$. An operation on X is a function from X^n to X.

Operations may be thought of as means of combining elements of a set to produce new elements of the set. The most common operations are binary operations (when $n = 2$).

EXAMPLE 1.18. $+$ and $\cdot$ are binary operations on $\mathbb{N}$. $-$ and $\div$ are not operations on $\mathbb{N}$.

EXAMPLE 1.19. Let $X = \mathbb{R}^3$, thought of as the set of 3-vectors. The function $x \mapsto -x$ is a unary operation on X, the function $(x, y) \mapsto x+y$ is a binary operation, and the function $(x, y, z) \mapsto x \times y \times z$ is a ternary operation.

If $f : X \to Y$, $g : X \to Y$, and $\star$ is a binary operation on Y, then there is a natural way to define a new function on X using $\star$. Define $f \star g$ by

$$f \star g : X \quad \to \quad Y$$
$$(f \star g)(x) \quad = \quad f(x) \star g(x).$$

EXAMPLE 1.20. Suppose f is the real function $f(x) = x^3$, and g is the real function $g(x) = 3x^2 - 1$. Then $f + g$ is the real function $x \mapsto x^3 + 3x^2 - 1$, and $f \cdot g$ is the real function $x \mapsto x^3(3x^2 - 1)$.

Another way to build new functions is by composition.

DEFINITION. Composition, $\circ$ Let $f : X \to Y$ and $g : Y \to Z$. Then the composition of g with f is the function,

$$g \circ f : X \quad \to \quad Z$$
$$x \quad \mapsto \quad g(f(x)).$$

EXAMPLE 1.21. Let f be the real function

$$f(x) = x^2.$$

Let g be the real function

$$g(x) = \sqrt{x}.$$

Then

$$(g \circ f)(x) = |x|.$$

What is $f \circ g$? (Be careful about the domain.)

EXAMPLE 1.22. Let

$$f : \mathbb{R} \rightarrow \mathbb{R}$$
$$x \mapsto 2x + 1$$

and let

$$g : \mathbb{R}^2 \rightarrow \mathbb{R}$$
$$(x, y) \mapsto x^2 + 3y^2.$$

Then

$$f \circ g : \mathbb{R}^2 \rightarrow \mathbb{R}$$
$$(x, y) \mapsto 2x^2 + 6y^2 + 1.$$

The function $g \circ f$ is not defined (why?).

1.4 Injections, Surjections, and Bijections

Among the most basic characteristics a function may have are the properties of injectivity, surjectivity, and bijectivity.

DEFINITION. Injection, One-to-one Let $f : X \rightarrow Y$. The function f is called an injection if, whenever x and y are distinct elements of X, we have $f(x) \neq f(y)$. Injections are also called one-to-one functions.

Another way of stating the definition (the contrapositive) is that if $f(x) = f(y)$, then $x = y$.

EXAMPLE 1.23. The real function $f(x) = x^3$ is an injection. To see this, let x and y be real numbers, and suppose that

$$f(x) = x^3 = y^3 = f(y).$$

Then

$$x = (x^3)^{1/3} = (y^3)^{1/3} = y.$$

So, for $x, y \in X$,

$$f(x) = f(y) \text{ only if } x = y.$$

EXAMPLE 1.24. The real function $f(x) = x^2$ is not an injection since

$$f(2) = 4 = f(-2).$$

Observe that a single example suffices to show that f is not an injection.

EXAMPLE 1.25. Suppose $f : X \to Y$ and $g : Y \to Z$. Prove that if f and g are injective, so is $g \circ f$.

PROOF. Suppose that $g \circ f(x) = g \circ f(y)$. Since g is injective, this means that $f(x) = f(y)$. Since f is injective, this in turn means that $x = y$. Therefore $g \circ f$ is injective, as desired.   □

(See Exercise 1.20.)

DEFINITION. Surjection, Onto   Let $f : X \to Y$. We say f is a surjection from X to Y if $\text{Ran}(f) = Y$. We also describe this by saying that f is onto Y.

EXAMPLE 1.26. The function $f : \mathbb{R} \to \mathbb{R}$ defined by $f(x) = x^2$ is not a surjection. For instance, -1 is in the codomain of f, but $-1 \notin \text{Ran}(f)$. Therefore, $\text{Ran}(f) \subsetneq \mathbb{R}$.

EXAMPLE 1.27. Let $Y = \{x \in \mathbb{R} \mid x \geq 0\}$, and let $f : \mathbb{R} \to Y$ be given by $f(x) = x^2$. Then f is a surjection. To prove this, we need to show that $Y = \text{Ran}(f)$. We know that $\text{Ran}(f) \subseteq Y$, so we must show $Y \subseteq \text{Ran}(f)$. Let $y \in Y$, so y is a nonnegative real number. Then $\sqrt{y} \in \mathbb{R}$, and $f(\sqrt{y}) = y$. So $y \in \text{Ran}(f)$. Since y was an arbitrary element of Y, $Y \subseteq \text{Ran}(f)$. Hence $Y = \text{Ran}(f)$ and f is a surjection.

Whether a function is a surjection depends on the choice of the codomain. A function is always onto its range. You might wonder why one would not simply define the codomain as the range of the function (guaranteeing that the function is a surjection). One reason is that we may be more interested in relating two sets using functions than we are in any particular function between the sets. We study an important application of functions to relating sets in Chapter 6, where we use functions to compare the size of sets. This is of particular interest when comparing infinite sets and has led to deep insights in the foundations of mathematics.

If we put the ideas of an injection and a surjection together, we arrive at the key idea of a bijection.

DEFINITION. Bijection, $\rightarrowtail$ Let $f : X \to Y$. If f is an injection and a surjection, then f is a bijection. This is written as $f : X \rightarrowtail Y$.

Why are bijections so important? From a theoretical point of view, functions may be used to relate the domain and the codomain of the function. If you are familiar with one set, you may be able to develop insights into a different set by finding a function between the sets that preserves some of the key characteristics of the sets. For instance, an injection can "interpret" one set into a different set. If the injection preserves the critical information from the domain, we can behave as if the domain of the function is virtually a subset of the codomain by using the function to "rename" the elements of the domain. If the function is a bijection, and it preserves key structural features of the domain, we can treat the domain and the codomain as virtually the same set. What the key structural features are depends on the area of mathematics you are studying. For example, if you are studying algebraic structures, you are probably most interested in preserving the operations of the structure. If you are studying geometry, you are interested in functions that preserve shape. The preservation of key structural features of the domain or codomain often allows us to translate knowledge of one set into equivalent knowledge of another set.

DEFINITION. Permutation Let X be a set. A permutation of X is a bijection $f : X \rightarrowtail X$.

EXAMPLE 1.28. Let $f : \mathbb{Z} \to \mathbb{Z}$ be defined by

$$f(x) \; = \; x + 1.$$

Then f is a permutation of $\mathbb{Z}$.

EXAMPLE 1.29. Let $X = \{0, 1, -1\}$. Then $f : X \to X$ given by $f(x) = -x$ is a permutation of X.

1.5 Images and Inverses

Functions can be used to define subsets of given sets.

DEFINITION. Image, $f[\ ]$ Let $f : X \to Y$ and $W \subseteq X$. The image of W under f, written $f[W]$, is the set

$$\{f(x) \mid x \in W\}.$$

So, if $f : X \to Y$, then

$$\text{Ran}(f) \; = \; f[X].$$

EXAMPLE 1.30. Suppose f is the real function $f(x) = x^2 + 3$. Let $W = \{-2, 2, 3\}$ and $Z = (-1, 2)$. Then $f[W] = \{7, 12\}$ and $f[Z] = [3, 7)$.

In applications of mathematics, functions often describe numerical relationships between measurable observations. So if $f : X \to Y$ and $a \in X$, then $f(a)$ is the predicted or actual measurement associated with a. In this context, one is often interested in determining which elements of X are associated with a value, b, in the codomain of f.

DEFINITION 1.31. Inverse image, Pre-image, $f^{-1}(\ )$ Let $f : X \to Y$ and $b \in Y$. Then the inverse image of b under f, $f^{-1}(b)$, is the set

$$\{x \in X \mid f(x) = b\}.$$

This set is also called the pre-image of b under f.

Note that if $b \notin \mathrm{Ran}(f)$, then $f^{-1}(b) = \emptyset$. If f is an injection, then for any $b \in \mathrm{Ran}(f)$, $f^{-1}(b)$ has a single element.

DEFINITION. Inverse image, Pre-image, $f^{-1}[\ ]$ Let $f : X \rightarrow Y$ and $Z \subseteq Y$. The inverse image of Z under f, or the pre-image of Z under f, is the set

$$f^{-1}[Z] \ = \ \{x \in X \ | \ f(x) \in Z\}.$$

We use $f^{-1}[\ ]$ to mean the inverse image of a *subset* of the codomain and $f^{-1}(\ )$ to mean the inverse image of an *element* of the codomain— both are subsets of the domain of f. If $Z \cap \mathrm{Ran}(f) \ = \ \emptyset$, then

$$f^{-1}[Z] \ = \ \emptyset.$$

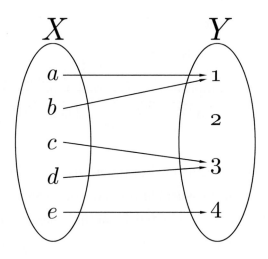

FIGURE 1.1 Picture of f

EXAMPLE 1.32. Let f be as in Figure 1.1 Then $f[\{b, c\}] = \{1, 3\}$ and $f^{-1}[\{1, 3\}] = \{a, b, c, d\}$.

EXAMPLE 1.33. Let g be the real function $g(x) = x^2 + 3$. If $b \in \mathbb{R}$ and $b > 3$, then

$$g^{-1}(b) = \{\sqrt{b-3}, \ -\sqrt{b-3}\}.$$

If $b = 3$, then $g^{-1}(3) = \{0\}$. If $b < 3$, then $g^{-1}(b)$ is empty.

EXAMPLE 1.34. Let h be the real function $h(x) = e^x$. If $b \in \mathbb{R}$ and $b > 0$, then

$$h^{-1}(b) = \{\log_e(b)\}.$$

For instance,

$$h^{-1}(1) = \{0\}.$$

Because h is strictly increasing, the inverse image of any element of the codomain ($\mathbb{R}$) is either a set with a single element or the empty set.

Let $I = (a, b)$, where $a, b \in \mathbb{R}$ and $0 < a < b$ (that is, I is the open interval with end points a and b). Then

$$h^{-1}[I] = (\log_e(a), \log_e(b)).$$

We have discussed the construction of new functions from existing functions using algebraic operations and composition of functions. Another tool for building new functions from known functions is the inverse function.

DEFINITION 1.35. Inverse function Let $f : X \rightarrowtail Y$ be a bijection. Then the inverse function of f, $f^{-1} : Y \rightarrow X$, is the function with graph

$$\{(b, a) \in Y \times X \ | \ (a, b) \in \text{graph}(f)\}.$$

The function f^{-1} is defined by "reversing the arrows." For this to make sense, $f : X \rightarrow Y$ must be bijective. Indeed, if f were not surjective, then there would be an element y of Y that is not in the range of f and so cannot be mapped back to anything in X. If f were not injective, there would be elements z of Y that were the image of distinct elements x_1 and x_2 in X. One could not define $f^{-1}(z)$ without specifying how to choose a particular pre-image. Both these problems can be fixed. If f is injective but not surjective, one can define $g : X \rightarrowtail \mathrm{Ran}(f)$ by

$$g(x) \;=\; f(x)$$

for all $x \in X$. Then $g^{-1} : \mathrm{Ran}(f) \rightarrowtail X$. If f is not injective, the problem is trickier; but if one can find some subset of X on which f is injective, one could restrict one's attention to that set.

EXAMPLE 1.36. Let f be the real function $f(x) = x^2$. The function f is not an bijection, so it does not have an inverse function. However, the function

$$g : [0, \infty) \;\rightarrow\; [0, \infty)$$
$$x \;\mapsto\; x^2$$

is a bijection (see Fig. 1.2). In this case,

$$g^{-1}(y) \;=\; \sqrt{y}.$$

EXAMPLE 1.37. Let f be the real function $f(x) = e^x$. You know from calculus that f is an injection and that $\mathrm{Ran}(f) = \mathbb{R}^+$. Hence f is not a surjection, since the implicit codomain of a real function is $\mathbb{R}$. The function

$$g : \mathbb{R} \;\rightarrow\; \mathbb{R}^+$$
$$x \;\mapsto\; e^x$$

is a bijection and

$$g^{-1}(x) \;=\; \log_e(x).$$

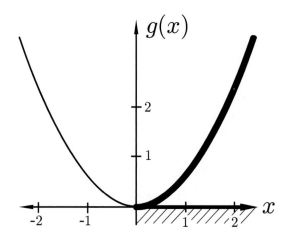

FIGURE 1.2 Picture of g

Warning: For $f : X \rightarrowtail Y$ to be a bijection, we have assigned two different meanings to $f^{-1}(b)$. In Definition 1.31, it means the set of points in X that get mapped to b. In Definition 1.35, it means the inverse function, f^{-1}, of the bijection f applied to the point $b \in Y$. However, if f is a bijection, so that the second definition makes sense, then these definitions are closely related. Suppose $a \in \text{Dom}(f)$ and $f(a) = b$. According to Definition 1.31, $f^{-1}(b) = \{a\}$, and by Definition 1.35, $f^{-1}(b) = a$. In practice the context will make clear which definition is intended.

DEFINITION. Identity function, $\text{id}|_X$ Let X be a set. The identity function on X, $\text{id}|_X : X \rightarrowtail X$, is the function defined by

$$\text{id}|_X(x) = x.$$

If $f : X \rightarrow Y$ is a bijection, then f^{-1} is the unique function such that

$$f^{-1} \circ f = \text{id}|_X$$

and

$$f \circ f^{-1} = \mathrm{id}|_Y.$$

Because $f(x) = x^2$ is not an injection, it has no inverse, even after restricting the codomain to be the range. Therefore in order to "invert" f, we considered a different function $g(x)$, which was equal to f on a subset of the domain of f and was an injection. In Example 1.36, we accomplished this by defining the function $g(x) = x^2$ with domain $\{x \in \mathbb{R} \mid x \geq 0\}$. Many of the functions that we need to invert for practical and theoretical reasons happen not to be injections and hence do not have inverse functions. One way to address this obstacle is to consider the function on a smaller domain.

Given a function $f : X \to Y$, we may wish to define an "inverse" of f on some subset of $W \subseteq X$ for which the *restriction* of f to W is an injection.

DEFINITION. Restricted domain, $f|_W$ Let $f : X \to Y$ and $W \subseteq X$. The restriction of f to W, written $f|_W$, is the function

$$f|_W : W \quad \to \quad Y$$
$$x \quad \mapsto \quad f(x).$$

So if $f : X \to Y$ and $W \subseteq X$, then

$$\mathrm{graph}(f|_W) = [W \times Y] \cap [\mathrm{graph}(f)].$$

EXAMPLE 1.38. Let $f(x) = (x-2)^4$. Let $W = [2, \infty)$. Then

$$f|_W : W \quad \to \quad [0, \infty)$$

is a bijection.

EXAMPLE 1.39. Let f be the real function $f(x) = \tan(x)$. Then

$$\mathrm{Dom}(f) = \{x \in \mathbb{R} \mid x \neq \pi/2 + k\pi, \ k \in \mathbb{Z}\}$$

and
$$\mathrm{Ran}(f) \;=\; \mathbb{R}.$$
The function f is periodic with period π and is therefore not an injection. Nonetheless, it is important to answer the question

At what angle(s), x, does $\tan(x)$ equal a particular value, $a \in \mathbb{R}$?

This is mathematically equivalent to asking

What is $\arctan(a)$?

In calculus this need was met by restricting the domain to a largest interval, I, such that
$$f|_I \;:\; I \rightarrowtail \mathbb{R}.$$
For any $k \in \mathbb{Z}$,
$$\left(\frac{(2k+1)\pi}{2}, \frac{(2k+3)\pi}{2} \right)$$
is such an interval. In order to define a specific function, the simplest of these intervals is selected, and we define
$$\mathrm{Tan} \;: = \; \tan|_{(-\pi/2,\pi/2)}.$$
Observe that
$$\mathrm{Tan} : (-\pi/2, \pi/2) \rightarrowtail \mathbb{R}.$$
So the function is invertible, that is, Tan has an inverse function,
$$\mathrm{Arctan} \;=\; \mathrm{Tan}^{-1}.$$

1.6 Sequences

In calculus we think of a *sequence* as a (possibly infinite) list of objects. We shall expand on that idea somewhat and express it in the language of functions.

DEFINITION. Finite sequence, $\langle a_n \mid n < N \rangle$ A finite sequence is a function f with domain $\ulcorner N \urcorner$, where $N \in \mathbb{N}$. We often identify the sequence with the ordered finite set $\langle a_n \mid n < N \rangle$, where $a_n = f(n)$, for $0 \leq n < N$.

This interpretation of a sequence as a type of function is easily extended to infinite sequences.

DEFINITION. Infinite sequence, $\langle a_n \mid n \in \mathbb{N} \rangle$ An infinite sequence is a function f with domain $\mathbb{N}$. We often identify the sequence with the ordered infinite set $\langle a_n \mid n \in \mathbb{N} \rangle$, where $a_n = f(n)$, for $n \in \mathbb{N}$.

REMARK. Interval in $\mathbb{Z}$ Actually, the word "sequence" is normally used to mean any function whose domain is an interval in $\mathbb{Z}$, where an *interval in* $\mathbb{Z}$ is the intersection of some real interval with $\mathbb{Z}$. For convenience in this book, we usually assume that the first element of any sequence is indexed by 0 or 1.

EXAMPLE 1.40. The sequence $\langle 0, 1, 4, 9, \ldots \rangle$ is given by the function $f(n) = n^2$.

The sequence $\langle 1, -1, 2, -2, 3, -3, \ldots \rangle$ is given by the function

$$f(n) = \begin{cases} \frac{n}{2} + 1, & n \text{ even} \\ -\frac{n+1}{2}, & n \text{ odd.} \end{cases}$$

Sequences can take values in any set (the codomain of the function f that defines the sequence). We talk of a *real sequence* if the values are real numbers, an *integer sequence* if they are all integers, and so on. It will turn out later that sequences with values in the two-element set $\{0, 1\}$ occur quite frequently, so we have a special name for them: we call them *binary sequences*.

DEFINITION. Binary sequence A finite binary sequence is a function $f : \ulcorner N \urcorner \to \ulcorner 2 \urcorner$ for some $N \in \mathbb{N}$. An infinite binary sequence is a function $f : \mathbb{N} \to \ulcorner 2 \urcorner$.

We often use the expression $\langle a_n \rangle$ for the sequence $\langle a_n \mid n \in \mathbb{N} \rangle$.

Functions are also used to "index" sets in order to build more complicated sets with generalized set operations. We discussed the union (or intersection) of more than two sets. You might ask whether it is possible to form unions or intersections of a large (infinite) collection of sets. There are two concerns that should be addressed in answering this question. We must be sure that the definition of the union of infinitely many sets is precise, that is, that it uniquely characterizes an object in the mathematical universe. We also need notation for managing this idea—how do we specify the sets over which we are taking the union?

DEFINITION. Infinite union, Index set, $\bigcup_{n=1}^{\infty} X_n$ For $n \in \mathbb{N}^+$, let X_n be a set. Then

$$\bigcup_{n=1}^{\infty} X_n = \{x \mid \text{for some } n \in \mathbb{N}^+, \ x \in X_n\}.$$

The set $\mathbb{N}^+$ is called the index set for the union.

This may be written in a few different ways.

NOTATION. $\bigcup_{n \in \mathbb{N}^+} X_n$ The following three expressions are all equal:

$$X_1 \cup X_2 \cup \dots \cup X_n \cup \dots$$

$$\bigcup_{n=1}^{\infty} X_n$$

$$\bigcup_{n \in \mathbb{N}^+} X_n.$$

We can use index sets other than $\mathbb{N}^+$.

DEFINITION. Family of sets, Indexed union, $\bigcup_{\alpha \in A} X_\alpha$ Let A be a set, and for $\alpha \in A$, let X_α be a set. The set

$$\mathcal{F} = \{X_\alpha \mid \alpha \in A\}$$

is called a family of sets indexed by A. Then

$$\bigcup_{\alpha \in A} X_\alpha = \{x \mid x \in X_\alpha \text{ for some } \alpha \in A\}.$$

The notation $\bigcup_{\alpha \in A} X_\alpha$ is read "the union over alpha in A of the sets X sub alpha."

So

$$x \in \bigcup_{\alpha \in A} X_\alpha \text{ if } x \in X_\alpha \text{ for some } \alpha \in A.$$

General intersections over a family of sets are defined analogously:

$$\bigcap_{\alpha \in A} X_\alpha = \{x \mid x \in X_\alpha \text{ for all } \alpha \in A\}.$$

EXAMPLE 1.41. Let $X_n = \{n + 1, n + 2, \ldots, 2n\}$ for each $n \in \mathbb{N}^+$.
Then

$$\bigcup_{n=1}^{\infty} X_n = \{k \in \mathbb{N} \mid k \geq 2\}$$

$$\bigcap_{n=1}^{\infty} X_n = \emptyset.$$

EXAMPLE 1.42. For each positive real number t, let $Y_t = [11/t, t]$.
Then

$$\bigcup_{t \in (\sqrt{11}, \infty)} Y_t = \mathbb{R}^+$$

$$\bigcap_{t \in [\sqrt{11}, \infty)} Y_t = \{\sqrt{11}\}.$$

EXAMPLE 1.43. Let $f : X \to Y$, $A \subseteq X$, and $B \subseteq Y$. Then

$$\bigcup_{a \in A} \{f(a)\} = f[A]$$

and

$$\bigcup_{b \in B} f^{-1}(b) = f^{-1}[B].$$

1.7 Russell's Paradox

As the ideas for set theory were explored, there were attempts to define
sets as broadly as possible. It was hoped that any collection of mathematical objects that could be defined by a formula would qualify as
a set. This belief was known as the General Comprehension Principle

(GCP) . Unfortunately, the GCP gave rise to conclusions that were unacceptable for mathematics.

Consider the collection defined by the following simple formula:

$$V = \{x \mid x \text{ is a set and } x = x\}.$$

If V is considered as a set, then since $V = V$,

$$V \in V.$$

If this is not an inconsistency, it is at least unsettling. Unfortunately, it gets worse. Consider the collection

$$X = \{x \mid x \notin x\}.$$

Then

$$X \in X \text{ if and only if } X \notin X.$$

This latter example is called Russell's paradox and shows that the GCP is false. Clearly there would have to be some control over which definitions give rise to sets. Axiomatic set theory was developed to provide rules for rigorously defining sets. We give a brief discussion in Appendix B.

1.8 Exercises

EXERCISE 1.1. Show that

$$\{n \in \mathbb{N} \mid n \text{ is odd and } n = k(k+1) \text{ for some } k \in \mathbb{N}\}$$

is empty.

EXERCISE 1.2. Let X and Y be subsets of some set U. Prove de Morgan's laws:

$$(X \cup Y)^c = X^c \cap Y^c$$
$$(X \cap Y)^c = X^c \cup Y^c.$$

EXERCISE 1.3. Let X, Y, and Z be sets. Prove

$$X \cap (Y \cup Z) = (X \cap Y) \cup (X \cap Z)$$
$$X \cup (Y \cap Z) = (X \cup Y) \cap (X \cup Z).$$

EXERCISE 1.4. Let $X = \ulcorner 2 \urcorner$, $Y = \ulcorner 3 \urcorner$, and $Z = \ulcorner 1 \urcorner$. What are the following sets:

  (i) $X \times Y$
  (ii) $X \times Y \times Z$
  (iii) $X \times Y \times Z \times \emptyset$
  (iv) $X \times X$
  (v) X^n

EXERCISE 1.5. Suppose X is a set with m elements and Y is a set with n elements. How many elements does $X \times Y$ have? Is the answer the same if one or both of the sets is empty?

EXERCISE 1.6. How many elements does $\emptyset \times \mathbb{N}$ have?

EXERCISE 1.7. Describe all possible intervals in $\mathbb{Z}$.

EXERCISE 1.8. Let X and Y be finite nonempty sets with m and n elements, respectively. How many functions are there from X to Y? How many injections? How many surjections? How many bijections?

EXERCISE 1.9. What happens in Exercise 1.8 if m or n is zero?

EXERCISE 1.10. For each of the following sets, which of the operations (addition, subtraction, multiplication, division, and exponentiation) are operations on the set:

  (i) $\mathbb{N}$
  (ii) $\mathbb{Z}$
  (iii) $\mathbb{Q}$
  (iv) $\mathbb{R}$

(v) $\mathbb{R}^+$

EXERCISE 1.11. Let f and g be real functions $f(x) = 3x + 8$ and $g(x) = x^2 - 5x$. What are $f \circ g$ and $g \circ f$? Is $(f \circ g) \circ f = f \circ (g \circ f)$?

EXERCISE 1.12. Write down all permutations of $\{a, b, c\}$.

EXERCISE 1.13. What is the natural generalization of Exercise 1.2 to an arbitrary number of sets? Verify your generalized laws.

EXERCISE 1.14. What is the natural generalization of Exercise 1.3 to an arbitrary number of sets? Verify your generalized laws.

EXERCISE 1.15. Let X be the set of all triangles in the plane, Y the set of all right-angled triangles, and Z the set of all nonisosceles triangles. For any triangle T, let $f(T)$ be the longest side of T and $g(T)$ be the maximum of the lengths of the sides of T. On which of the sets X, Y, Z is f a function? On which is g a function? What is the complement of Z in X? What is $Y \cap Z^c$?

EXERCISE 1.16. For each positive real t, let $X_t = (-t, t)$ and $Y_t = [-t, t]$. Describe

(i) $\bigcup_{t>0} X_t$ and $\bigcup_{t>0} Y_t$

(ii) $\bigcup_{0<t<10} X_t$ and $\bigcup_{0<t<10} Y_t$

(iii) $\bigcup_{0<t\leq 10} X_t$ and $\bigcup_{0<t\leq 10} Y_t$

(iv) $\bigcap_{t\geq 10} X_t$ and $\bigcap_{t\geq 10} Y_t$

(v) $\bigcap_{t>10} X_t$ and $\bigcap_{t>10} Y_t$

(vi) $\bigcap_{t>0} X_t$ and $\bigcap_{t>0} Y_t$

EXERCISE 1.17. Let f be the real function cosine, and let g be the real function $g(x) = \dfrac{x^2 + 1}{x^2 - 1}$.

(i) What are $f \circ g$, $g \circ f$, $f \circ f$, $g \circ g$, and $g \circ g \circ f$?

(ii) What are the domains and ranges of the real functions f, g, $f \circ g$, and $g \circ f$?

EXERCISE 1.18. Let X be the set of vertices of a square in the plane. How many permutations of X are there? How many of these come from rotations? How many come from reflections in lines? How many come from the composition of a rotation and a reflection?

EXERCISE 1.19. Which of the following real functions are injective and which are surjective:

(i) $f_1(x) = x^3 - x + 2$.

(ii) $f_2(x) = x^3 + x + 2$.

(iii) $f_3(x) = \dfrac{x^2 + 1}{x^2 - 1}$.

(iv) $f_4(x) = \begin{cases} -x^2 & x \le 0 \\ 2x + 3 & x > 0. \end{cases}$

EXERCISE 1.20. Suppose $f : X \to Y$ and $g : Y \to Z$. Prove that if $g \circ f$ is injective, then f is injective. Give an example to show that g need not be injective.

EXERCISE 1.21. Suppose $f : X \to Y$ and $g : Y \to Z$.

(i) Show that if f and g are surjective, so is $g \circ f$.
(ii) Show that if $g \circ f$ is surjective, then one of the two functions f, g must be surjective (which one?). Give an example to show that the other function need not be surjective.

EXERCISE 1.22. For what $n \in \mathbb{N}$ is the function $f(x) = x^n$ an injection?

EXERCISE 1.23. Let $f : \mathbb{R} \to \mathbb{R}$ be a polynomial of degree $n \in \mathbb{N}$. For what values of n must f be a surjection, and for what values is it not a surjection?

EXERCISE 1.24. Write down a bijection from $(X \times Y) \times Z$ to $X \times (Y \times Z)$. Prove that it is one-to-one and onto.

EXERCISE 1.25. Let X be a set with n elements. How many permutations of X are there?

EXERCISE 1.26. Let $f : \mathbb{R} \to \mathbb{R}$ be a function built using only natural numbers and addition, multiplication, and exponentiation (for instance f could be defined as $x \mapsto (x+3)^{x^2}$). What can you say about $f[\mathbb{N}]$? What can you say if we include subtraction or division?

EXERCISE 1.27. Let $f(x) = x^3 - x$. Find sets X and Y such that $f : X \to Y$ is a bijection. Is there a maximal choice of X? If there is, is it unique? Is there a maximal choice of y? If there is, is it unique?

EXERCISE 1.28. Let $f(x) = \tan(x)$. Use set notation to define the domain and range of f. What is $f^{-1}(1)$? What is $f^{-1}[\mathbb{R}^+]$?

EXERCISE 1.29. For each of the following real functions, find an interval X that contains more than one point and such that the function is a bijection from X to $f[X]$. Find a formula for the inverse function.

   (i) $f_1(x) = x^2 + 5x + 6$.
   (ii) $f_2(x) = x^3 - x + 2$.
   (iii) $f_3(x) = \dfrac{x^2 + 1}{x^2 - 1}$.
   (iv) $f_4(x) = \begin{cases} -x^2 & x \le 0 \\ 2x + 3 & x > 0 \end{cases}$.

EXERCISE 1.30. Find formulas for the following sequences:

   (i) $\langle 1, 2, 9, 28, 65, 126, \ldots \rangle$
   (ii) $\langle 1, -1, 1, -1, 1, -1, \ldots \rangle$
   (iii) $\langle 2, 1, 10, 27, 67, 125, 218, \ldots \rangle$
   (iv) $\langle 1, 1, 2, 3, 5, 8, 13, 21, \ldots \rangle$

EXERCISE 1.31. Let the real function f be strictly increasing. Show that for any $b \in \mathbb{R}$, $f^{-1}(b)$ is either empty or consists of a single element and that f is therefore an injection. If f is also a bijection, is the inverse function of f also strictly increasing?

EXERCISE 1.32. Let f be a real function that is a bijection. Show that the graph of f^{-1} is the reflection of the graph of f in the line $y = x$.

EXERCISE 1.33. Let $X_n = \{n+1, n+2, \ldots, 2n\}$ for each $n \in \mathbb{N}^+$ as in Example 1.41. What are

(i) $\bigcup_{n=1}^{5} X_n$
(ii) $\bigcap_{n=4}^{6} X_n$
(iii) $\bigcap_{k=1}^{5} \left[\bigcup_{n=1}^{k} X_n \right]$
(iv) $\bigcap_{k=5}^{\infty} \left[\bigcup_{n=3}^{k} X_n \right]$

EXERCISE 1.34. Verify the assertions of Example 1.42.

EXERCISE 1.35. Let $f : X \to Y$ and assume that $U_\alpha \subseteq X$ for every $\alpha \in A$, and $V_\beta \subseteq Y$ for every $\beta \in B$. Prove

(i) $\quad f\left(\bigcup_{\alpha \in A} U_\alpha \right) = \bigcup_{\alpha \in A} f(U_\alpha).$

(ii) $\quad f\left(\bigcap_{\alpha \in A} U_\alpha \right) \subseteq \bigcap_{\alpha \in A} f(U_\alpha).$

(iii) $\quad f^{-1}\left(\bigcup_{\beta \in B} V_\beta \right) = \bigcup_{\beta \in B} f^{-1}(V_\beta).$

(iv) $\quad f^{-1}\left(\bigcap_{\beta \in B} V_\beta \right) = \bigcap_{\beta \in B} f^{-1}(V_\beta).$

Note that (ii) has containment instead of equality. Give an example of proper containment in (ii). Find a condition on f that would ensure equality in (ii).

1.9 Hints to Get Started on Some Exercises

Exercise 1.2. You could do this with a Venn diagram. However, once there are more than three sets (see Exercise 1.13), this approach will be difficult. An algebraic proof will generalize more easily, so try to find one here. Argue for the two inclusions

$$(X \cup Y)^c \subseteq X^c \cap Y^c$$
$$X^c \cap Y^c \subseteq (X \cup Y)^c$$

separately. In the first one, for example, assume that $x \in (X \cup Y)^c$ and show that it must be in both X^c and Y^c.

Exercise 1.13. Part of the problem here is notation—what if you have more sets than letters? Start with a finite number of sets contained in U, and call them $X_1, \ldots, X_n$. What do you think the complement of their union is? Prove it as you did when $n = 2$ in Exercise 1.2. (See the advantage of having a proof in Exercise 1.2 that did not use Venn diagrams? One of the reasons mathematicians like to have multiple proofs of the same theorem is that each proof is likely to generalize in a different way.)

Can you make the same argument work if your sets are indexed by some infinite index set?

Now do the same thing with the complement of the intersection.

Exercise 1.14. Again there is a notational problem, but while Y and Z play the same rôle in Exercise 1.3, X plays a different rôle. So rewrite the equations as

$$X \cap (Y_1 \cup Y_2) = (X \cap Y_1) \cup (X \cap Y_2)$$
$$X \cup (Y_1 \cap Y_2) = (X \cup Y_1) \cap (X \cup Y_2)$$

and see if you can generalize these.

Exercise 1.35. (i) Again, this reduces to proving two containments. If y is in the left-hand side, then there must be some x_0 in some U_{α_0} such that $f(x) = y$. But then y is in $f(U_{\alpha_0})$, so y is in the right-hand side.

Conversely, if y is in the right-hand side, then it must be in $f(U_{\alpha_0})$ for some $\alpha_0 \in A$. But then y is in $f\left(\cup_{\alpha \in A} U_\alpha\right)$ and so is in the left-hand side.

CHAPTER 2

Relations

2.1 Definitions

DEFINITION. Relation Let X and Y be sets. A relation from X to Y is a subset of $X \times Y$.

Alternatively, any set of ordered pairs is a relation. If $Y = X$, we say that R is a relation on X.

NOTATION. xRy Let X and Y be sets and R be a relation on $X \times Y$. If $x \in X$ and $y \in Y$, then we may express that x bears relation R to y (that is $(x, y) \in R$) by writing xRy.

So for X and Y sets, $x \in X$, $y \in Y$, and R a relation on $X \times Y$,

$$xRy \quad \text{if and only if} \quad (x, y) \in R.$$

EXAMPLE 2.1. Let $\leq$ be the usual ordering on $\mathbb{Q}$. Then $\leq$ is a relation on $\mathbb{Q}$. We write

$$1/2 \leq 2$$

to express that $1/2$ bears the relation $\leq$ to 2.

EXAMPLE 2.2. Define a relation R from $\mathbb{Z}$ to $\mathbb{R}$ by xRy if $x > y+3$. Then we could write $7 \ R \ \sqrt{2}$ or $(7, \sqrt{2}) \in R$ to say that $(7, \sqrt{2})$ is in the relation.

EXAMPLE 2.3. Let $X = \{2, 7, 17, 27, 35, 72\}$. Define a relation R by xRy if $x \neq y$ and x and y have a digit in common. Then

$$R = \{(2, 27), (2, 72), (7, 17), (7, 27), (7, 72), (17, 7), (17, 27), (17, 72),$$
$$(27, 2), (27, 7), (27, 17), (27, 72), (72, 2), (72, 7), (72, 17), (72, 27)\}.$$

EXAMPLE 2.4. Let P be the set of all polygons in the plane. Define a relation E by saying $(x, y) \in E$ if x and y have the same number of sides.

How do mathematicians use relations? A relation on a set can be used to impose structure. In Example 2.1, the usual ordering relation $\leq$ on $\mathbb{Q}$ allows us to think of rational numbers as lying on a number line, which provides additional insight into rational numbers. In Example 2.4, we can use the relation to break polygons up into the sets of triangles, quadrilaterals, pentagons, and so on.

A function $f : X \to Y$ can be thought of as a very special sort of relation from X to Y. Indeed, the graph of the function is a set of ordered pairs in $X \times Y$, but it has the additional property that every x in X occurs exactly once as a first element of a pair in the relation. As we discussed in Section 1.3, functions are a useful way to relate sets.

Let X be a set, and let R be a relation on X. Here are some important properties the relation may or may not have.

DEFINITION.   Reflexive R is reflexive if for every $x \in X$,

$$xRx.$$

Symmetric R is symmetric if for any $x, y \in X$,

$$xRy \text{ implies } yRx.$$

Antisymmetric R is antisymmetric if for any $x, y \in X$,

$$[(x, y) \in R \text{ and } (y, x) \in R] \text{ implies } x = y.$$

Transitive R is transitive if for any $x, y, z \in X$,

$$[xRy \text{ and } yRz] \text{ implies } [xRz].$$

Which of these four properties applies to the relations given in Examples 2.1–2.4 (Exercise 2.1)?

2.2 Orderings

A relation on a set may be thought of as part of the structure imposed on the set. Among the most important relations on a set are order relations.

DEFINITION. Partial ordering Let X be a set and R a relation on X. We say that R is a partial ordering (or partial order) if

(1) R is reflexive
(2) R is antisymmetric
(3) R is transitive

EXAMPLE 2.5. Let X be a family of sets. The relation $\subseteq$ is a partial ordering on X. Every set is a subset of itself, so the relation is reflexive. If $Y \subseteq Z$ and $Z \subseteq Y$, then $Y = Z$, so the relation is antisymmetric. Finally, if $Y \subseteq Z$ and $Z \subseteq W$, then $Y \subseteq W$, so the relation is transitive.

EXAMPLE 2.6. Let R be the relation on $\mathbb{N}^+$ defined by xRy if and only if there is $z \in \mathbb{N}^+$ such that

$$xz = y.$$

Then R is a partial ordering of $\mathbb{N}^+$. (Prove this in Exercise 2.2.)

DEFINITION. Linear ordering Let X be a set and R be a partial ordering of X. We say that R is a linear ordering, also called a *total ordering*, provided that, for any $x, y \in X$, either xRy or yRx.

Note that since a linear ordering is antisymmetric, for any distinct x and y, exactly one of xRy and yRx holds.

EXAMPLE 2.7. The ordering $\leq$ on $\mathbb{N}$ (or $\mathbb{R}$) is a linear ordering. So is the relation $\geq$. The relation $<$ is not (why?).

EXAMPLE 2.8. Let $X = \mathbb{R}^n$. We can define a reflexive relation on X as follows. Let $x = (a_1, \ldots, a_n)$ and $y = (b_1, \ldots, b_n)$ be distinct members of X. Let $k \in \mathbb{N}^+$ be the least number such that $a_k \neq b_k$. Then we define

$$xRy \quad \text{if and only if } a_k < b_k.$$

Then R is a linear ordering of X. It is called the *dictionary ordering*.

The notion of a linear ordering is probably natural for you, and you have used it intuitively since you began studying arithmetic. The relation $\leq$ helps you to visualize the set as a line in which the relative location of two elements of the set is determined by the linear ordering. If you are considering a set with operations, this in turn can help in visualizing how operations behave. For instance, think of using a number line to visualize addition, subtraction, and multiplication of integers.

Partial orderings are generalizations of linear orderings, and $\leq$ is the most obvious example of a linear ordering. Because of this, the normal symbol for a partial ordering is $\preceq$ (it is also reminiscent of the symbol $\subseteq$, which is the example most mathematicians keep in mind when thinking about a partial ordering).

EXAMPLE 2.9. Let X be the set of all collections of apples and oranges. If x, y are in X, then say $x \preceq y$ if the number of apples in x is less than or equal to the number of apples in y and the number of oranges in x is less than or equal to the number of oranges in y. This is a partial ordering. You may not be able to compare apples to oranges, but you can say that two apples and five oranges is inferior to four apples and six oranges!

One way to visualize a partial order $\preceq$ on a finite set X is to imagine arrows connecting distinct elements of X, x and y, if $x \preceq y$ and there is no third distinct point z satisfying $x \preceq z \preceq y$. Then two elements a and b in X will satisfy $a \preceq b$ if and only if you can get from a to b by following a path of arrows.

EXAMPLE 2.10. Consider the graph on the set $X = \{a, b, c, d, e, f\}$ given in Figure 2.1.

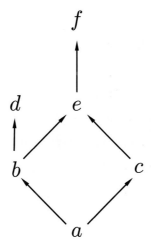

FIGURE 2.1 Picture of a partial order

It illustrates the partial order that could be described as the smallest reflexive and transitive relation $\preceq$ on X that satisfies $a \preceq b$, $a \preceq c$, $b \preceq d$, $b \preceq e$, $c \preceq e$, $e \preceq f$.

2.3 Equivalence Relations

DEFINITION. Equivalence relation Let X be a set and R a relation on X. We say R is an equivalence relation if

(1) R is reflexive
(2) R is symmetric
(3) R is transitive

EXAMPLE 2.11. Define a relation R on $\mathbb{R}$ by xRy if and only if $x^2 = y^2$. Then R is an equivalence relation.

EXAMPLE 2.12. Let R be a relation defined on $\mathbb{Z} \times \mathbb{Z}$ as follows. If $a, b, c, d \in \mathbb{Z}$,

$$(a, b) \ R \ (c, d) \ \text{ if and only if } \ a + d \ = \ b + c. \qquad (2.13)$$

Then R is an equivalence relation. Indeed, let us check the three properties.

(1) Reflexive: By (2.13), we have $(a, b) \ R \ (a, b)$ if $a + b = a + b$, which clearly holds.

(2) Symmetric: Suppose $(a, b) \ R \ (c, d)$, so $a + d \ = \ b + c$. To see if $(c, d) \ R \ (a, b)$, we must check whether $c + b \ = \ d + a$; but this holds by the commutativity of addition.

(3) Transitive: Suppose $(a, b) \ R \ (c, d)$ and $(c, d) \ R \ (e, f)$. We must check that $(a, b) \ R \ (e, f)$, in other words that

$$a + f \ = \ b + e. \qquad (2.14)$$

We have $a + d \ = \ b + c$ and $c + f \ = \ d + e$, and adding these two equations we get

$$a + d + c + f \ = \ b + c + d + e. \qquad (2.15)$$

Cancelling $c + d$ from each side of (2.15), we get (2.14) as desired.

EXAMPLE 2.16. Let R be a relation on $X \ = \ \mathbb{Z} \times \mathbb{N}^{+}$ defined by

$$(a, b) \ R \ (c, d) \ \text{ if and only if } \ ad \ = \ bc.$$

Then R is an equivalence relation on X. (Prove this in Exercise 2.4.)

EXAMPLE 2.17. Let $f : X \to Y$. Define a relation R_f on X by

$$x \, R_f \, y \text{ if and only if } f(x) \, = \, f(y).$$

Then R_f is an equivalence relation. We check the conditions for an equivalence relation:

- R_f is clearly reflexive since, for any $x \in X$,

$$f(x) \, = \, f(x).$$

- R_f is symmetric since, for any $x \in X$ and $y \in X$,

$$f(x) \, = \, f(y) \text{ if and only if } f(y) \, = \, f(x).$$

- To show R_f is transitive, let $x, y, z \in X$. If $f(x) \, = \, f(y)$ and $f(y) \, = \, f(z)$, then $f(x) \, = \, f(z)$.

Equivalence relations have three of the key properties of identity. They allow us to relate objects in a set that we wish to consider as "the same" in a given context. This allows us to focus on which differences between mathematical objects are relevant to the discussion at hand, and which are not. For this reason, a common symbol for an equivalence relation is $\sim$.

DEFINITION. Equivalence class, $[x]_R$ Let R be an equivalence relation on a set X. If $x \in X$, then the equivalence class of x modulo R, denoted by $[x]_R$, is

$$[x]_R \, = \, \{y \in X \mid x R y\}.$$

If $y \in [x]_R$, we call y a representative element of $[x]_R$. The set of all equivalence classes $\{[x]_R \mid x \in X\}$ is written X/R. It is called the quotient space of X by R.

We may use $[x]$ for the equivalence class of x provided that the equivalence relation is clear.

NOTATION. Equivalence mod R, $\equiv_R$, $\sim$ Let R be an equivalence relation on a set X. We may express that xRy by writing

$$x \equiv y \bmod R$$

or

$$x \equiv_R y$$

or

$$x \sim y.$$

PROPOSITION 2.18. *Suppose that $\sim$ is an equivalence relation on* X. *Let $x, y \in X$. If $x \sim y$, then*

$$[x] \;=\; [y]. \tag{2.19}$$

If x is not equivalent to y $(x \nsim y)$, then

$$[x] \cap [y] \;=\; \emptyset.$$

PROOF. (i) Assume $x \sim y$. Let us show that $[x] \subseteq [y]$. Let $z \in [x]$. This means that $x \sim z$. Since $\sim$ is symmetric and $x \sim y$, we have $y \sim x$. As $y \sim x$ and $x \sim z$, by transitivity of $\sim$ we get that $y \sim z$. Therefore $z \in [y]$. Since z is an arbitrary element of $[x]$, we have shown that $[x] \subseteq [y]$.

As $y \sim x$, the same argument with x and y swapped gives $[y] \subseteq [x]$, and therefore $[x] \;=\; [y]$.

(ii) Now assume that x and y are not equivalent. We must show that there is no z such that $z \in [x]$ and $z \in [y]$. We will argue by contradiction. Suppose there were such a z. Then we would have

$$x \sim z \qquad \text{and} \qquad y \sim z.$$

By symmetry, we have also that $z \sim y$, and by transitivity, we then have that $x \sim y$. This contradicts the assumption that x is not equivalent to y. So if x and y are not equivalent, no z can exist that is simultaneously in both $[x]$ and $[y]$. Therefore $[x]$ and $[y]$ are disjoint sets, as required. $\square$

So what have we shown? We have not shown that any particular relation is an equivalence relation. Rather we have shown that any equivalence relation on a set partitions the set into disjoint equivalence classes.

As we shall see throughout this book, and you will see throughout your mathematical studies, this is a surprisingly powerful tool.

DEFINITION. Pairwise disjoint Let $\{X_\alpha \mid \alpha \in A\}$ be a family of sets. The family is pairwise disjoint if for any $\alpha, \beta \in A$, $\alpha \neq \beta$,

$$X_\alpha \cap X_\beta = \emptyset.$$

DEFINITION. Partition Let Y be a set and $\mathcal{F} = \{X_\alpha \mid \alpha \in A\}$ be a family of nonempty sets. The collection $\mathcal{F}$ is a partition of Y if $\mathcal{F}$ is pairwise disjoint and

$$Y = \bigcup_{\alpha \in A} X_\alpha.$$

Given an equivalence relation $\sim$ on a set X, the equivalence classes with respect to $\sim$ give a partition of X. Conversely, partitions give rise to equivalence relations.

THEOREM 2.20.   (i) *Let X be a set, and $\sim$ an equivalence relation on X. Then $X/\sim$ is a partition of X.*

(ii) *Conversely, let $\{X_\alpha \mid \alpha \in A\}$ be a partition of X. Let $\sim$ be the relation on X defined by $x \sim y$ whenever x and y are members of the same set in the partition. Then $\sim$ is an equivalence relation.*

PROOF. Part (i) of the theorem is Proposition 2.18 restated, and we gave the proof above. To prove the converse, we must show that the relation $\sim$ defined as in part (ii) of the theorem is an equivalence.

- Reflexivity: Let $x \in X$. Then x is in some X_{α_0}, as the union of all these sets is all of X. Therefore $x \sim x$.

- Symmetry: Suppose $x \sim y$. Then there is some X_{α_0} such that $x \in X_{\alpha_0}$ and $y \in X_{\alpha_0}$. This implies that $y \sim x$.

- Transitivity: Suppose $x \sim y$ and $y \sim z$. Then there are sets X_{α_0} and X_{α_1} such that both x and y are in X_{α_0}, and both y and z are in X_{α_1}. But since the sets X_α form a partition, and y is in both X_{α_0} and X_{α_1}, we must have that $X_{\alpha_0} = X_{\alpha_1}$. This implies that x and z are in the same member of the partition, and so $x \sim z$.

$\square$

2.4 Constructing Bijections

Let us consider a particularly interesting and important abstract application of equivalence classes. Let $f : X \to Y$. The function f need not be an injection or surjection. However, we have already discussed the desirability of finding an "inverse" for f, even when it fails to meet the necessary conditions for the existence of an inverse. In Section 1.3 we considered the function $f|_D$, where $D \subseteq X$ and $f|_D$ is an injection. Another approach is to use the function f to create a new function on a distinct domain that preserves much of the information of f.

We use f to induce an equivalence relation on X. Define a relation $\sim$ on X by

$$x \sim y \text{ if and only if } f(x) = f(y).$$

We showed in Example 2.17 that $\sim$ is an equivalence relation; it is the equivalence relation on X induced by f. The equivalence relation $\sim$ induces a partition of X, namely $X/\sim$ (which is the set $\{[x] \mid x \in X\}$ of all equivalence classes).

NOTATION. X/f Let $f : X \to Y$ and $\sim$ be the equivalence relation on X induced by f. We write X/f for the set of equivalence classes induced by $\sim$ on X.

An equivalence class in X/f is the inverse image of an element in $\mathrm{Ran}(f)$. That is, if $x \in X$ and $f(x) = y$,

$$[x] = f^{-1}(y).$$

So

$$X/f = \{f^{-1}(y) \mid y \in \mathrm{Ran}(f)\}.$$

The elements of X/f are called the level sets of f. The inspiration for this comes from thinking of a topographical map. The curves on a topographical map corresponding to fixed altitudes are called level curves. Consider the function from a point on a map to the altitude of the physical location represented by the point on the map. Level curves on the map are subsets of the level sets of this function.

NOTATION. Π_f Let $f : X \to Y$. The function $\Pi_f : X \to X/f$ is defined by $\Pi_f(x) = [x]_f$, where $[x]_f$ is the equivalence class of x with respect to the equivalence relation induced by f.

Let $Z \subseteq Y$ be the range of f. We define a new function, $\widehat{f} : X/f \to Z$ by

$$\widehat{f}([x]) = f(x).$$

The function $\widehat{f}$ is closely related to f; in fact, for every $x \in X$,

$$f(x) = \widehat{f} \circ \Pi_f(x).$$

This is sometimes illustrated with a diagram, as in Figure 2.2.

The function $\widehat{f}$ is a bijection. In this sense, every function can be canonically associated with a bijection. We consider the function that we looked at in Section 1.3.

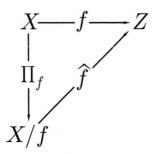

FIGURE 2.2 Making a function into a bijection

EXAMPLE 2.21. Let $f(x) = \tan(x)$. As we discussed earlier, we can "invert" this function by considering the function Tan : $(-\pi/2, \pi/2) \to \mathbb{R}$ by

$$\text{Tan} = \tan|_{(-\pi/2, \pi/2)}.$$

The function Tan is a bijection and has an inverse,

$$\text{Arctan} : \mathbb{R} \to (-\pi/2, \pi/2).$$

For any $k \in \mathbb{Z}$ there is a corresponding restriction of tan,

$$\tan\big|_{\left(\frac{(2k+1)\pi}{2}, \frac{(2k+3)\pi}{2}\right)}$$

which is a bijection and therefore has an inverse function.

Another bijection can be constructed on the equivalence classes induced by $f(x) = \tan(x)$. A level set of f is $[x]_f = \{x + k\pi \mid k \in \mathbb{Z}\}$. Let X be the domain of tan. Then

$$X/f = \{[x]_f \mid x \in X\}.$$

We can interpret an equivalence class $[x]_f$ with respect to angles in standard position in the Cartesian plane. The equivalence class of x is the set of angles in standard position that have terminal side collinear with the terminal side of the angle x—see Figure 2.3.

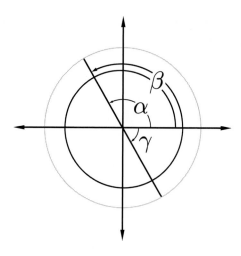

FIGURE 2.3 Collinear angles

Following the construction outlined above, the function $\Pi_f : X \to X/f$ is the function

$$\Pi_f(x) \;=\; [x]_f \;=\; \{x + k\pi \mid k \in \mathbb{Z}\}.$$

The function $\widehat{f} : X/f \to \mathbb{R}$ given by

$$\widehat{f}([x]_f) \;=\; f(x)$$

is a bijection. Furthermore,

$$\tan \;=\; \widehat{f} \circ \Pi_f.$$

If $x \in X$, then $\Pi_f(x)$ is the set of all angles that have terminal side collinear with the terminal side of angle x in standard position. Thus Π_f tells us that tan can distinguish only the slope of the terminal side of the angle—not the quadrant of the angle or how many revolutions the angle subtended.

2.5 Modular Arithmetic

We define an equivalence relation that will help us derive insights in number theory.

DEFINITION.   Divides, $a \mid b$ Let a and b be integers. Then a divides b, written $a \mid b$, if there is $c \in \mathbb{Z}$ such that

$$a \cdot c \; = \; b.$$

DEFINITION.   Congruence, $x \equiv y \mod n$, $\equiv_n$ Let $x, y, n \in \mathbb{Z}$ and $n > 1$. Then

$$x \equiv y \mod n$$

(or $x \equiv_n y$) if

$$n \mid (x - y).$$

The relation $\equiv_n$ on $\mathbb{Z}$ is called congruence mod n.

THEOREM 2.22. *Congruence mod n is an equivalence relation on* $\mathbb{Z}$.

Exercise 2.5: Prove Theorem 2.22.

DEFINITION.   Congruence class   The equivalence classes of the relation $\equiv_n$ are called congruence classes, residue classes, or remainder classes mod n. The set of congruence classes mod n can be written $\mathbb{Z}_n$ or $\mathbb{Z}/n\mathbb{Z}$.

Of course $\mathbb{Z}_n$ is a partition of $\mathbb{Z}$. When $n = 2$, the residue classes are called the even and the odd numbers. Many of the facts you know about even and odd numbers generalize if you think of them as residue classes. What are the residue classes for $n = 3$?

We leave it as an exercise (Exercise 2.6) to prove that two integers are in the same remainder class mod n provided that they have the same remainder when divided by n.

NOTATION.   [a]   Fix a natural number $n \geq 2$. Let a be in $\mathbb{Z}$. We represent the equivalence class of a with respect to the relation $\equiv_n$ by $[a]$.

PROPOSITION 2.23. *If $a \equiv r \mod n$ and $b \equiv s \mod n$, then*

$$\text{(i)} \qquad a + b \equiv r + s \mod n$$

and

$$\text{(ii)} \qquad ab \equiv rs \mod n.$$

PROOF. (i) Assume that $a \equiv r \mod n$ and $b \equiv s \mod n$. Then $n|(a - r)$ and $n|(b - s)$. So

$$n|(a + b - (r + s)).$$

Therefore

$$a + b \equiv r + s \mod n$$

proving (i).

To prove (ii), note that there are $i, j \in \mathbb{Z}$ such that

$$a = ni + r$$

and

$$b = nj + s.$$

Then

$$ab = n^2 ji + rnj + sni + rs = n(nji + rj + si) + rs.$$

Therefore

$$n|(ab - rs)$$

and

$$ab \equiv rs \mod n.$$

$\square$

Hence the algebraic operations that $\mathbb{Z}_n$ "inherits" from $\mathbb{Z}$ are well-defined. That is, we may define $+$ and $\cdot$ on $\mathbb{Z}_n$ by

$$[a] + [b] = [a + b] \qquad (2.24)$$

and

$$[a] \cdot [b] = [a \cdot b]. \qquad (2.25)$$

In mathematics, when you ask whether something is "well-defined," you mean that somewhere in the definition a choice was made, and you want to know whether a different choice would have resulted in the same final result. For example, let $X_1 = \{-2, 2\}$ and let $X_2 = \{-1, 2\}$. Define y_1 by "Choose x in X_1 and let $y_1 = x^2$." Define y_2 by "Choose x in X_2 and let $y_2 = x^2$." Then y_1 is well-defined and is the number 4; but y_2 is not well-defined, as different choices of x give rise to different numbers.

In (2.24) and (2.25), the right-hand sides depend a priori on a particular choice of elements from the equivalence classes $[a]$ and $[b]$. But Proposition 2.23 ensures that sum and product so defined are independent of the choice of representatives of the equivalence classes.

EXAMPLE 2.26. In Z_2, addition and multiplication are defined as follows:

(1) $[0] + [0] = [0]$.
(2) $[0] + [1] = [1] + [0] = [1]$.
(3) $[1] + [1] = [0]$.
(4) $[0] \cdot [0] = [0] \cdot [1] = [1] \cdot [0] = [0]$.
(5) $[1] \cdot [1] = [1]$.

Notice that if you read $[0]$ as "even" and $[1]$ as "odd," these are rules that you learned a long time ago.

When working with modular arithmetic we may pick the representatives of remainder classes that best suit our needs. For instance,

$$79 \cdot 23 \equiv 2 \cdot 2 \equiv 4 \mod 7.$$

In other words,

$$[79 \cdot 23] = [79] \cdot [23] = [2] \cdot [2] = [4].$$

EXAMPLE 2.27. You may recall from your early exposure to multiplication tables that multiplication by 9 results in a product whose digits sum to 9. This generalizes nicely with modular arithmetic. Specifically, if $a_n \in \ulcorner 10 \urcorner$ for $0 \leq n \leq N$, then

$$\sum_{n=0}^{N} a_n 10^n \equiv \sum_{n=0}^{N} a_n \mod 9. \tag{2.28}$$

The remainder of any integer divided by 9 equals the remainder of the sum of the digits of that integer when divided by 9.

PROOF. The key observation is that

$$10 \equiv 1 \mod 9.$$

Therefore

$$10^2 \equiv 1 \cdot 1 \equiv 1 \qquad \mod 9$$
$$10^3 \equiv 1 \cdot 1 \cdot 1 \equiv 1 \qquad \mod 9,$$

and so on for any power of 10:

$$10^n \equiv 1 \mod 9 \quad \text{for all } n \in \mathbb{N}.$$

(This induction to all powers of 10 is straightforward, but to prove it formally we shall need the notion of mathematical induction from Chapter 4). Therefore on the left-hand side of (2.28), working mod 9, we can replace all the powers of 10 by 1, and this gives us the right-hand side. □

EXAMPLE 2.29. The observation that a number's residue mod 9 is the same as that of the sum of the digits can be used in a technique called "casting out nines" to check arithmetic.

For example, consider the following (incorrect) sum. The number in the penultimate column is the sum of the digits, and the number in the last column is the repeated sum of the digits until reaching a number between 0 and 9.

1588	22	4
+1805	14	5
3493	19	1

If the addition had been correctly performed, the remainder mod 9 of the sum would equal the sum of the remainders; so we know a mistake was made.

EXAMPLE 2.30. What is the last digit of 7^7?

We want to know $7^7 \mod 10$. Note that modulo 10, $7^0 \equiv 1$, $7^1 \equiv 7$, $7^2 \equiv 9$, $7^3 \equiv 3$, $7^4 \equiv 1$. So $7^7 = 7^4 7^3 \equiv 1 \cdot 3 \equiv 3$, and so 3 is the last digit of 7^7.

EXAMPLE 2.31. What is the last digit of 7^{7^7}?

By Example 2.30, we see that the residues of $7^n \mod 10$ repeat themselves every time n increases by 4. Therefore if $m \equiv n \mod 4$, then $7^m \equiv 7^n \mod 10$.

What is $7^7 \mod 4$? Well $7^1 \equiv 3 \mod 4$, $7^2 \equiv 1 \mod 4$, so $7^7 \equiv (7^2)^3 \cdot 7 \equiv 3 \mod 4$. Therefore

$$7^{7^7} \equiv 7^3 \equiv 3 \mod 10.$$

2.6 Exercises

EXERCISE 2.1. Which of the properties of reflexivity, symmetry, antisymmetry, and transitivity apply to the relations given in Examples 2.1–2.4?

EXERCISE 2.2. Prove that the relation in Example 2.6 is a partial ordering.

EXERCISE 2.3. List every pair in the relation given in Example 2.10.

EXERCISE 2.4. Prove that the relation in Example 2.16 is an equivalence.

EXERCISE 2.5. Prove that congruence mod n is an equivalence relation on $\mathbb{Z}$.

EXERCISE 2.6. Prove that two integers are in the same congruence class mod n if and only if they have the same remainder when divided by n.

EXERCISE 2.7. Suppose R is a relation on X. What does it mean if R is both a partial order and an equivalence?

EXERCISE 2.8. Consider the relations on people "is a brother of," "is a sibling of," "is a parent of," "is married to," and "is a descendant of." Which of the properties of reflexivity, symmetry, antisymmetry, and transitivity do each of these relations have?

EXERCISE 2.9. Let $X = \{k \in \mathbb{N} : k \geq 2\}$. Consider the following relations on X:

(i) j R_1 k if and only if $\gcd(j, k) > 1$ (gcd stands for *greatest common divisor*).

(ii) j R_2 k if and only if j and k are coprime (i.e., $\gcd(j, k) = 1$).

(iii) j R_3 k if and only if $j|k$.

(iv) j R_4 k if and only if

$$\{p : p \text{ is prime and } p|j\} = \{q : q \text{ is prime and } q|k\}.$$

For each relation, say which of the properties of reflexivity, symmetry, antisymmetry, and transitivity it has.

EXERCISE 2.10. For j, k in $\mathbb{N}^+$, define two relations R_1 and R_2 by jR_1k if j and k have a digit in common (but not necessarily in the same place) and j R_2 k if j and k have a common digit in the same place (so, for example, 108 R_1 82, but $(108, 82) \notin R_2$).

(i) If $j = \sum_{m=0}^{M} a_m 10^m$ and $k = \sum_{n=0}^{N} b_n 10^n$, with $a_M \neq 0$ and $b_N \neq 0$, how can one mathematically define R_1 and R_2 in terms of the coefficients a_m and b_n?

(ii) Which of the four properties of reflexivity, symmetry, antisymmetry, and transitivity do R_1 and R_2 have?

EXERCISE 2.11. Let $X = \{a, b\}$. List all possible relations on X, and say which are reflexive, which are symmetric, which are antisymmetric, and which are transitive.

EXERCISE 2.12. How many relations are there on a set with three elements? How many of these are reflexive? How many are symmetric? How many are antisymmetric?

EXERCISE 2.13. Repeat Exercise 2.12 for a set with N elements.

EXERCISE 2.14. The sum of two even integers is even, the sum of an even and an odd integer is odd, and the sum of two odd integers is even. What is the generalization of this statement to residue classes mod 3?

EXERCISE 2.15. What is the last digit of 3^{5^7}? Of 7^{5^3}? Of 11^{10^6}? Of 8^{5^4}?

EXERCISE 2.16. What is $2^{1000000}$ mod 17? What is 17^{77} mod 14?

EXERCISE 2.17. Show that a number's residue mod 3 is the same as the sum of its digits.

EXERCISE 2.18. Show that the assertion of Exercise 2.17 is not true mod n for any value of n except 3 and 9.

EXERCISE 2.19. Prove that there are an infinite number of natural numbers that cannot be written as the sum of three squares. (Hint: Look at the possible residues mod 8).

EXERCISE 2.20. Let $f : X \to Y$ and $g : Y \to Z$. What can you say about the relationship between X/f and $X/(g \circ f)$?

EXERCISE 2.21. Let R be the relation on $X = \mathbb{Z} \times \mathbb{N}^+$ defined in Example 2.16. Define an operation $\star$ on X/R as follows: for $x = (a, b)$ and $y = (c, d)$,

$$[x] \star [y] = [(ad + bc, cd)].$$

Is $\star$ well-defined?

EXERCISE 2.22. Let X be the set of functions from finite subsets of $\mathbb{N}$ to $\ulcorner 2 \urcorner$ (that is $f \in X$ if and only if there is a finite set $D \subseteq \mathbb{N}$ such that $f : D \to \ulcorner 2 \urcorner$). Define a relation R on X as follows: if $f, g \in X$, fRg if and only if $\mathrm{Dom}(g) \subseteq \mathrm{Dom}(f)$ and $g = f|_{\mathrm{Dom}(g)}$. Is R a partial ordering? Is R an equivalence relation?

EXERCISE 2.23. Let X be the set of all infinite binary sequences. Define a relation R on X as follows: For any $f, g \in X$, fRg if and only if $f^{-1}(1) \subseteq g^{-1}(1)$. Is R a partial ordering? Is R an equivalence relation?

EXERCISE 2.24. Let $X = \{\ulcorner n \urcorner \mid n \in \mathbb{N}\}$. Let R be a relation on X defined by $x, y \in X$ if and only if $x \subseteq y$. Prove that R is a linear ordering.

EXERCISE 2.25. Let $X = \{f : \mathbb{R} \to \mathbb{R} \mid f \text{ is a surjection}\}$. Define a relation R on X by fRg if and only if $f(0) = g(0)$. Prove that R is an equivalence relation. Let $F : X \to \mathbb{R}$ be defined by $F(f) = f(0)$. Show that the level sets of F are the equivalence classes of X/R. That is, show that

$$X/R = X/F.$$

EXERCISE 2.26. Let $f : X \to Y$. Show that X/f is composed of singletons (sets with exactly one element) if and only if f is an injection.

CHAPTER 3

Proofs

3.1 Mathematics and Proofs

The primary activity of research mathematicians is proving mathematical claims. Depending on the depth of the claim, the relationship of the claim to other mathematical claims, and various other factors, a mathematical statement that has been proved is generally called a *theorem, proposition, corollary,* or *lemma.* A mathematical statement that has not been proved, but that is expected to be true, is commonly called a *conjecture.* A statement that is accepted as a starting point for arguments without being proved is called an *axiom.*

Some mathematical results are so fundamental, deep, difficult, surprising, or otherwise noteworthy that they are named. Part of your initiation as a member of the community of mathematicians is becoming familiar with some of these named statements—and we shall prove a few of them in this book.

It is likely that most of the mathematics you have studied has been the application of theorems to deriving solutions of relatively concrete problems. Here we begin learning how to prove theorems. Most students find the transition from computational mathematics to mathematical proofs very challenging.

3.1.1 What Is a Mathematical Proof? The nature of a mathematical proof depends on the context. There is a formal notion of a mathematical proof:

> A finite sequence of formal mathematical statements such that each statement either (a) is an axiom or assumption or (b) follows by formal rules of logical deduction from previous statements in the sequence.

Most mathematicians do not think of mathematical proofs as formal mathematical proofs, and virtually no mathematician writes formal mathematical proofs. This is because a formal proof is a hopelessly cumbersome thing and is generally outside the scope of human capability, even for the most elementary mathematical statements. Rather, mathematicians write proofs that are sequences of statements in a combination of natural language and formal mathematical symbols (interspersed with diagrams, questions, references, and other devices that are intended to assist the reader in understanding the proof) that can be thought of as representing a purely formal argument. A good practical definition of a mathematical proof is

> An argument in favor of a mathematical statement that will convince the preponderance of knowledgeable mathematicians of the truth of the mathematical statement.

This definition is somewhat imprecise, and mathematicians can disagree on whether an argument is a proof, particularly for extremely difficult or deep arguments. However, given some time for careful consideration, the mathematical community nearly always reaches a unanimous consensus on whether a mathematical argument is a proof.

The notion of a mathematical proof for the student is similar to the general idea of a mathematical proof. The differences are due to the type of statement that the student is proving and the reasons for requesting that the student prove the statement. The statements that you will be proving are known to professional mathematicians or can be proved with relatively little effort by your instructors. Clearly the statements you will be proving require different conditions for a satisfactory proof than those stated above for the professional mathematician. Let us define a successful argument by the student as follows: An argument for a mathematical statement is one that

- the instructor can understand
- the instructor cannot refute
- uses only assumptions that the instructor considers admissible

Note that refuting an argument is not the same as refuting the original claim. The sentence "The square of every real number is nonnegative because all real numbers are nonnegative" is a false proof of a true statement. The sentence "The square of every real number is nonnegative because all triangles have three sides" fails the first test: while both statements are true, your instructor will not see how the first follows from the second.

In this book, the solutions to the problems will be an exposition in natural language enhanced by mathematical expressions. The student is expected to learn the conventions of mathematical grammar and argument and use them. Like most conventions, these are often determined by tradition or precedent. It can be quite difficult, initially, to determine whether your mathematical exposition meets the standards of your instructor. Practice, with feedback from a reader experienced in reading mathematics, is the best way to develop good proof-writing skills. Remember, readers of mathematics are quite impatient with trying to decipher what the author *means* to say—mathematics is sufficiently challenging when the author writes precisely what he or she intends. Most of the burden of communication is on the author of a mathematical proof, not the reader. A proof can be logically correct but so difficult to follow that it is unacceptable to your instructor.

3.1.2 Why Proofs? Why are proofs the primary medium of mathematics? Mathematicians depend on proofs for certainty and explanation. Once a proof is accepted by the mathematical community, it is virtually unheard of that the result is subsequently refuted. This was not always the case: in the nineteenth century there were serious disputes as to whether results had really been proved or not (see Section 5.3 for an example; see also the book [4] for a very extensive treatment of the development of rigor in mathematical reasoning). This led to our modern notion of a "rigorous" mathematical argument. While one might argue that it is possible that in the twenty-first century a new standard of rigor will reject what we currently consider to be proofs, our current ideas have been stable for over a century, and most mathematicians (including the authors of this book) do not expect that there will be a philosophical shift.

For very complicated results, writing a detailed proof helps the author convince himself or herself of the truth of the claim. After a mathematician has hit upon the key idea behind an argument, there is a lot of hard work left developing the details of the argument. Many promising ideas fail as the author attempts to write a detailed argument based on the idea. Finally, proofs often provide a deeper insight into the result and the mathematical objects that are the subject of the proof. Indeed, even very clever proofs that fail to provide mathematical insights are held in lower regard, by some, than arguments that elucidate the topic.

Mathematical proofs are strongly related to formal proofs in a purely logical sense. It is supposed that the existence of an informal mathematical proof is overwhelming evidence for the existence of a formal mathematical proof. If it is not clear that the informal proof could conceivably be transformed into a formal argument, it is doubtful that the informal argument will be accepted by the mathematical community. Consequently, mathematical arguments have a transparent underlying logical structure.

For this reason we shall begin our discussion of mathematical proofs with a brief discussion of propositional logic. Despite its abstractness, the topic is straightforward, and most of the claims of this section may be confirmed with some careful, patient thinking.

3.2 Propositional Logic

Propositional logic studies how the truth or falsehood of compound statements is determined by the truth or falsehood of the constituent statements. It gives us a way of reliably deriving true conclusions from true assumptions.

DEFINITION. Truth value If P is a statement that is true, then P has truth value 1. If P is a statement that is false, P has truth value 0. We write $T(P)$ for the truth value of P.

Truth values can be thought of as a function $T : S \to \ulcorner 2 \urcorner$, where S is the set of all statements. When investigating the abstract principles of propositional logic, we consider possible assignments of truth values to variables representing statements. We are interested in claims that are independent of any particular assignment of truth values to the propositional variables. We use the integers 0 and 1 to represent truth values because it allows us to use arithmetic operations in propositional logic. Other authors prefer F and T.

DEFINITION. Propositional connectives The symbols $\land$, $\lor$, $\neg$, and $\Rightarrow$ are propositional connectives. They are defined as follows for statements P and Q.

Connective	Name	Definition
$\neg$	negation	$T(\neg P) = 1 - T(P)$
$\wedge$	conjunction	$T(P \wedge Q) = T(P) \cdot T(Q)$
$\vee$	disjunction	$T(P \vee Q) = T(P) + T(Q) - T(P) \cdot T(Q)$
$\Rightarrow$	implication	$T(P \Rightarrow Q) = 1 - T(P) + T(P) \cdot T(Q)$

In the expression "$P \Rightarrow Q$," the statement P is called the *antecedent* or *hypothesis* and Q is called the *consequence* or *conclusion*.

Propositional connectives are formal equivalents of natural language connectives.

Connective	Natural Language Equivalent
$\neg$	not
$\wedge$	and
$\vee$	or
$\Rightarrow$	implies

Check that the formulas defining the propositional connectives give the meaning that you anticipate. For example, check that the definition of the truth value for $P \wedge Q$ means that $P \wedge Q$ is true if and only if both P and Q are true.

Propositional connectives approximate natural language connectives. Propositional connectives are formal and precise, while natural language connectives are imprecise and somewhat more expressive— consequently the approximation is imperfect. We saw an example of this when contrasting mathematicians' use of the connective "or" with its use in everyday language. For precision in mathematics we interpret the connectives formally—even when using natural language expressions.

We can build very complicated compound statements by using logical connectives. Naturally, there are rules for building correct statements with connectives.

DEFINITION.   Atomic statement   An atomic statement is a statement with no explicit propositional connectives.

An atomic statement is usually represented by a capital letter.

DEFINITION.   Well-formed statement   We define a well-formed propositional statement recursively as follows:

- Atomic statements are well-formed.
- If P and Q are well-formed statements, then the following are
  well-formed statements:

$$(\neg P)$$
$$(P \wedge Q)$$
$$(P \vee Q)$$
$$(P \Rightarrow Q)$$

In practice the parentheses are dropped unless there is the poten-
tial for ambiguity. Additionally, "[" and "]" may be substituted for
parentheses in the interests of readability. For any assignment of truth
values to the atomic statements in a well-formed statement, the compound statement will have a well-defined truth value.

DEFINITION.   Compound statement   A compound statement is
a well-formed statement composed of atomic statements and propositional connectives.

3.2.1 Propositional Equivalence.
One purpose of propositional logic
is to give tools for assessing the truth of a compound statement without necessarily having to understand the specific meaning of the atomic
statements. That is, some statements are demonstrably true or false
by virtue of their form. Central to this understanding is the idea of
propositional equivalence.

DEFINITION.   Propositional equivalence   Let P and Q be well-
formed statements built from atomic statements. We say that P and

Q are propositionally equivalent provided that $T(P) = T(Q)$ for any assignment of truth values to the constituent atomic statements.

If P and Q are propositionally equivalent, we may write

$$P \equiv Q.$$

EXAMPLE 3.1.

$$[P \Rightarrow Q] \equiv [(\neg Q) \Rightarrow (\neg P)].$$

This is a very important example of a propositional equivalence. We will show this by considering all possible assignments of truth values to P and Q. Let us set this up in what is popularly called a *truth table*. We consider all possible assignments of truth values to P and Q, and we compare the truth values of the compound statements under consideration:

$T(P)$	$T(Q)$	$T(P \Rightarrow Q)$	$T((\neg Q) \Rightarrow (\neg(P)))$
0	0	1	1
0	1	1	1
1	0	0	0
1	1	1	1

Each row of the truth table represents a particular assignment of truth values to the atomic statements P and Q. The four possible assignments are exhausted by the rows of the truth table. The truth values of the compound statements agree in each row of the truth table, so the statements are equivalent.

EXAMPLE 3.2.

$$[\neg(P \wedge Q)] \equiv [(\neg P) \vee (\neg Q)] \tag{3.3}$$

$$[\neg(P \vee Q)] \equiv [(\neg P) \wedge (\neg Q)] \tag{3.4}$$

Statements (3.3) and (3.4) are known as de Morgan's laws. (How are they related to Exercise 1.2?)

With two possible exceptions, once you carefully study what these connectives mean, you should understand them intuitively. One exception is that the logical and mathematical "or," $\vee$, is inclusive. We discussed this in Section 1.1. The other exception is the logical connective "implies," $\Rightarrow$.

3.2.2 Implication. Students often find it confusing that the implication $P \Rightarrow Q$ can be true when the consequence, Q, is false. This is understandable when we consider that implications are usually employed in argument in the following syllogism:

$$P$$
$$P \Rightarrow Q$$

therefore,

$$Q$$

(i.e., if P is true, and $P \Rightarrow Q$, then Q is true). This syllogism is the most important rule of logical deduction (called *modus ponens*). Logical implication is so often used to demonstrate the truth of the consequence that it is easy to understand why one might mistakenly think that the consequence must follow from the implication rather than following from the *antecedent*. Consider the following statement:

If you are the king of France, then I am a monkey's uncle.

Is this statement true? Presumably you are not the king of France, and I do not believe that I am a monkey's uncle. So both the antecedent and the consequence are false. However the statement is true. In fact, this statement is logically equivalent to the statement:

If I am not a monkey's uncle, then you are not the king of France.

The definition of logical implication says that an implication in which the antecedent is false gives no information about the consequence.

Hence, any logical implication with the antecedent "You are the king of France" will be true.

There is an additional concern with logical implication. In natural language (and intuitively in mathematics), the statement

$$P \Rightarrow Q$$

suggests a relationship between the statements P and Q—namely that the truth of P somehow forces the truth of Q. As a propositional connective, this relationship between P and Q is not required for logical implication. The truth of $P \Rightarrow Q$ is a function of the truth values of P and Q, not of their *meanings*. In mathematical writing, it is understood that not only is the implication logically true but that P and Q are related and that the truth of P indeed forces the truth of Q. For instance, consider the statement

$$\mathbb{N} \subset \mathbb{Q} \Rightarrow 3 > 2.$$

This statement is true by the formal definition of $\Rightarrow$. In fact, as a propositional statement, we could replace the antecedent with any other statement, true or false, and the conditional statement would be true. However, such a statement is mathematically unacceptable since the antecedent and the consequence have nothing to do with each other. We are not concerned with the accidental truth values of atomic statements but with the mathematical connections between these statements, which comply with, yet go beyond, the formal definition of logical connectives.

3.2.3 Converse and Contrapositive. Most mathematical claims have the form of an implication. Therefore you need to be familiar with the conventional nomenclature surrounding logical implication. Suppose we are interested in a particular logical implication,

$$P \Rightarrow Q.$$

There are two other logical implications which are naturally associated with $P \Rightarrow Q$. One is the contrapositive,

$$\neg Q \Rightarrow \neg P.$$

An implication and its contrapositive are propositionally equivalent.

EXAMPLE 3.5. The statement

> If this is an insect then it has six legs

is propositionally equivalent to the statement

> If this does not have six legs, it is not an insect.

EXAMPLE 3.6. The contrapositive of

> A whale is a fish

is

> If it is not a fish then it is not a whale.

Example 3.6 illustrates that a statement need not be true in order to have a contrapositive (which is, of course, still propositionally equivalent to the original conditional statement). It also illustrates that conditional statements in natural language need not include the word "if" or "then," nor be written in a particular form, in order to be a conditional statement.

The converse of a conditional statement,

$$P \Rightarrow Q,$$

is the conditional statement

$$Q \Rightarrow P.$$

A conditional statement and its converse are not propositionally equivalent. You can easily check that $P \Rightarrow Q$ and $Q \Rightarrow P$ have different truth values if $T(P) = 1$ and $T(Q) = 0$.

EXAMPLE 3.7. What is the converse to the statement "All fish live in water"? Since this is written in natural language, there is no unique answer. An obvious converse is

If something lives in water, then it is a fish.

If we put together an implication and its converse, we get the biconditional connective.

DEFINITION. Biconditional, $\iff$ Let P and Q be statements. The biconditional, written $\iff$, is defined as follows.

Connective	Name	Definition
$\iff$	biconditional	$T(P \iff Q) = T(P \Rightarrow Q) \cdot T(Q \Rightarrow P)$

The biconditional connective is the formal interpretation of "if and only if." This phrase is so commonly used in mathematics that it has its own abbreviation: *iff*.

Other natural language words that can be translated into propositional connectives are "necessary" and "sufficient." The statement

In order for P to hold, it is necessary that Q holds

is equivalent to $P \Rightarrow Q$. The statement

In order for P to hold, it is sufficient that Q holds

is equivalent to $Q \Rightarrow P$. Combining these two, it follows that the statement

In order for P to hold, it is necessary and sufficient that Q holds

is equivalent to $P \iff Q$.

3.3 Formulas

Loosely speaking, a formula is a mathematical expression with variables. Corresponding to each variable, x_i, appearing in a formula is a universe, U_i, from which that variable may be substituted.

DEFINITION.   Open formula   An open mathematical formula in variables $x_1, \ldots, x_n$ is a mathematical expression in which substitution of the x_i $(1 \leq i \leq n)$ by specific elements from U_i yields a mathematical statement.

EXAMPLE 3.8. Consider the formula,

$$x^2 + y^2 \;=\; z^2$$

in variables x, y, and z, all with universe $\mathbb{N}$. Any substitution of the variables with natural numbers results in a statement. For instance,

$$3^2 + 4^2 \;=\; 5^2$$

or

$$1^2 + 1^2 \;=\; 2^2.$$

Of course, statements can be true or false, so some substitutions yield true statements, while others will yield false statements.

In discussing a general formula in n variables, we may use the notation $P(x_1, \ldots, x_n)$. For $1 \leq i \leq n$, let U_i be the universe of the variable x_i, and $a_i \in U_i$. The statement that results from the substitution of a_i for x_i, $1 \leq i \leq n$, is written $P(a_1, \ldots, a_n)$.

If $P(x_1, \ldots, x_n)$ is a formula in variables $x_1, \ldots, x_n$, and for $1 \leq i \leq n$, U_i is the universe of x_i, then we may think of $(x_1, \ldots, x_n)$ as a single variable with universe $U \;=\; \prod_{1 \leq i \leq n} U_i$.

Formulas can fulfill many purposes in mathematics:

(1) Characterize relationships between quantities.
(2) Define computations.
(3) Define sets.
(4) Define functions.

EXAMPLE 3.9. Consider an open formula, $P(x, y)$, in two variables,

$$x^2 + y^2 \;=\; 1,$$

with universe $\mathbb{R}^2$. That is, the universe of x is $\mathbb{R}$ and the universe of y is $\mathbb{R}$. One way to think of $P(x, y)$ is as a means to partition $\mathbb{R}^2$ into two sets:

(1) the subset of the Cartesian plane for which the equation is true, namely the unit circle

(2) the subset of the Cartesian plane for which the equation is false, the complement of the unit circle in $\mathbb{R}^2$

DEFINITION. Characteristic set, χ_P Let $P(x)$ be a formula and U the universe of the variable x. The subset of U for which the formula P holds is written χ_P. The set χ_P is called the characteristic set of $P(x)$.

So,

$$\chi_{\neg P} = U \setminus \chi_P.$$

3.3.1 Formulas and Propositional Connectives. Propositional logic is easily extended to formulas. Let $P(x)$ and $Q(x)$ be formulas in the variable x, with universe U. Let

$$R(x) = P(x) \wedge Q(x).$$

Then the characteristic set of $R(x)$ is given by

$$\chi_R = \{a \in U \mid T(\, P(a) \wedge Q(a)\,) = 1\,\}$$

Hence

$$\chi_R = \chi_P \cap \chi_Q.$$

The propositional connective $\wedge$ is strongly associated with the set operation $\cap$. Similarly $\vee$ may be associated with $\cup$, $\neg$ with complement (in U), and $\Rightarrow$ with $\subseteq$.

3.4 Quantifiers

Let $P(x)$ be a formula in one variable. If we substitute a constant, $a \in U$, for x we arrive at a statement $P(a)$. However, suppose that we are interested in $P(x)$ with regard to some set $X \subseteq U$ rather than a particular element of U. In particular, we ask if $P(a)$ is a true statement for all $a \in X$. Recall that one of the roles of a formula is to define sets. For any formula $P(x)$, universe U, and $X \subseteq U$, $P(x)$ partitions X into two sets—those elements of X for which P is true and those for which P is false. In this sense, asking whether P holds for all $x \in X$, or whether it holds for some $x \in X$ (which is complementary to asking whether $\neg P$ holds for all $x \in X$), is asking whether P defines a new or interesting subset of X.

Just as propositional connectives were introduced to formalize the linguistic behavior of certain widely employed natural language connectives (and, or, implies, not), we shall also formalize "quantification" over sets.

DEFINITION. Universal quantifier, $(\forall x \in X)\ P(x)$ Let $P(x)$ be a formula in one variable with universe U. Let $X \subseteq U$. Let Q be the statement

$$(\forall x \in X)P(x).$$

Then Q is true if for every $a \in X$, $P(a)$ is true. Otherwise Q is false.

The notation

$$(\forall x \in X)\ P(x)$$

is a shorthand for

$$(\forall x)\ ([x \in X] \Rightarrow [P(x)]).$$

The statement "$(\forall x \in X)\ P(x)$" is read "for all x in X, $P(x)$."
We have

$$(\forall x \in X)\ P(x) \iff X \subseteq \chi_P.$$

DEFINITION. Existential quantifier, $(\exists x \in X)\ P(x)$ Let $P(x)$ be a formula in one variable with universe U. Let $X \subseteq U$, $X \neq \emptyset$. Let Q be the statement

$$(\exists x \in X)\ P(x).$$

Then Q is true if there is some $a \in X$ for which $P(a)$ is true. Otherwise Q is false.

The expression

$$(\exists x \in X) \; P(x)$$

is a shorthand for

$$(\exists x) \; [(x \in X) \wedge P(x)].$$

The statement "$(\exists x \in X) \; P(x)$" is read "there exists x in X, such that $P(x)$." The quantifier "$\forall$" is the formal equivalent of the natural language expression "for all" or "every." The quantifier "$\exists$" is the formal equivalent of "for some" or "there exists ... such that ...."

Provided that the universe of a variable is clear, or not relevant to the discussion, it is common to suppress the universe in the expression of the statement. For instance, if $P(x)$ is a formula with universe U, we may write

$$(\forall x) \; P(x)$$

instead of

$$(\forall x \in U) \; P(x).$$

3.4.1 Multiple Quantifiers. Let $P(x_1, \ldots, x_n)$ be a formula in $n \geq 2$ variables. Then the formula

$$(\forall x_1) \; P(x_1, \ldots, x_n)$$

is a formula in the $n - 1$ variables $x_2, \ldots, x_n$. Similarly, the formula

$$(\exists x_1) \; P(x_1, \ldots, x_n)$$

is a formula in $n - 1$ variables.

EXAMPLE 3.10. Consider the formula in five variables

$$P(x, x_0, L, \varepsilon, \delta) \; := \; (0 < \mid x - x_0 \mid < \delta) \Rightarrow (\mid \sin(x) - L \mid < \varepsilon)$$

with all variables having universe $\mathbb{R}$.

Then $(\forall x_0)P(x, x_0, L, \varepsilon, \delta)$ is a formula in four variables, $(\forall x_0)(\exists L)$ $P(x, x_0, L, \varepsilon, \delta)$ is a formula in three variables, and

$$(\forall x_0)(\exists L)(\forall \varepsilon)P(x, x_0, L, \varepsilon, \delta)$$

is a formula in two variables.

DEFINITION.   Open variable, Bound variable   In the formula $P(x)$, x is an open variable. In the formulas

$$(\forall x)\ P(x), \quad (\exists x)\ P(x), \quad (\forall x)\ Q(x,y), \quad (\exists x)\ Q(x,y)$$

x is a bound or quantified variable, and in the last two, y is an open variable.

3.4.2 Quantifier Order. In the discussion below, we need to discuss quantifiers generically, that is without regard to whether the quantifier under discussion is universal or existential. So we shall introduce some convenient notation just for this section.

NOTATION. $(\mathcal{Q}x)\ P(x)$ We use the notation

$$(\mathcal{Q}x)\ P(x)$$

to generically represent

$$(\forall x)\ P(x)$$

and

$$(\exists x)\ P(x).$$

Let $\mathcal{Q}_1, \ldots, \mathcal{Q}_n$ be logical quantifiers and $P(x_1, \ldots, x_n)$ be a formula with open variables $x_1, \ldots, x_n$. Then

$$(\mathcal{Q}_1 x_1)(\mathcal{Q}_2 x_2)(\ldots)(\mathcal{Q}_n x_n)\ P(x_1, \ldots, x_n)$$

is a statement.

EXAMPLE 3.11. Consider a statement S in the form

$$S = (\forall x \in X) \, (\exists y \in Y) \, P(x, y).$$

S is true if for each $a \in X$,

$$(\exists y \in Y) \, P(a, y)$$

is true. This is satisfied provided that for each $a \in X$, there is an element of Y (let us call it b_a to remind us that this particular element of Y is associated with the previous choice, a) such that

$$P(a, b_a)$$

is true. So b_a is selected with a in mind. Statements in this form are especially important in mathematics because the definition of the limit in calculus is a statement in the form of this example.

Let us return to the statement

$$(\mathcal{Q}_1 x_1)(\ldots)(\mathcal{Q}_n x_n) \, P(x_1, \ldots, x_n).$$

The order of the quantifiers is significant. If $1 \le i < j \le n$, x_i behaves like a parameter from the point of view of x_j (that is, x_i is fixed from the point of view of x_j). Put another way, x_j is chosen with respect to the substitutions of $x_1, \ldots, x_{j-1}$ but without consideration for $x_{j+1}, \ldots, x_n$.

One always reads from the left. The statement

$$(\forall x_1)(\mathcal{Q}_2 x_2)(\ldots)(\mathcal{Q}_n x_n) \, P(x_1, \ldots, x_n)$$

is the same as

$$(\forall x_1) \, [(\mathcal{Q}_2 x_2)(\ldots)(\mathcal{Q}_n x_n) \, P(x_1, \ldots, x_n)],$$

or, in other words, for every choice of x_1, the statement

$$(\mathcal{Q}_2 x_2)(\ldots)(\mathcal{Q}_n x_n) \, P(x_1, \ldots, x_n)$$

is true. Similarly, the statement

$$(\exists x_1)(\mathcal{Q}_2 x_2)(\ldots)(\mathcal{Q}_n x_n) \ P(x_1, \ldots, x_n)$$

is the same as

$$(\exists x_1) \ [(\mathcal{Q}_2 x_2)(\ldots)(\mathcal{Q}_n x_n) \ P(x_1, \ldots, x_n)],$$

or, in other words, there is some choice of x_1 for which the statement

$$(\mathcal{Q}_2 x_2)(\ldots)(\mathcal{Q}_n x_n) \ P(x_1, \ldots, x_n)$$

about the $n - 1$ variables $x_2, \ldots, x_n$ is true.

EXAMPLE 3.12. Order of quantifiers is important, as you can see from the following:

$$(\forall x \in X) \ (\exists y \in Y) \ P(x, y)$$

is not equivalent to

$$(\exists y \in Y) \ (\forall x \in X) \ P(x, y).$$

For instance, the statement

$$(\forall x \in \mathbb{R}) \ (\exists y \in \mathbb{R}) \ (y \ = \ x^2)$$

is true. But

$$(\exists y \in \mathbb{R}) \ (\forall x \in \mathbb{R}) \ (y \ = \ x^2)$$

is false. The statement

$$[(\exists y \in Y) \ (\forall x \in X) \ P(x, y)] \Rightarrow [(\forall x \in X) \ (\exists y \in Y) \ P(x, y)]$$

is true. The converse clearly fails.

3.4.3 Negation of Quantifiers. In an important sense, $\wedge$ and $\vee$ are complementary. By de Morgan's laws (3.3) and (3.4), the negation of a simple conjunction is a disjunction of negations. Similarly, the negation of a simple disjunction is a conjunction of negations. Universal and existential quantifiers are also complementary. We observe that

$$[\neg(\forall x) \ P(x)] \equiv [(\exists x) \ \neg P(x)]$$

for any formula, $P(x)$. Similarly

$$[\neg(\exists x)\ P(x)] \equiv [(\forall x)\ \neg P(x)].$$

Of course, $P(x)$ itself may be a formula that has numerous quantifiers and bound variables. Let us suppose that

$$P(x) = (\exists y)\ Q(x, y). \tag{3.13}$$

Then the following statements are equivalent (for any choice of P and Q satisfying the identity (3.13)):

$$\neg(\forall x)\ P(x)$$

$$(\exists x)\ \neg P(x)$$

$$\neg(\forall x)\ (\exists y)\ Q(x, y)$$

$$(\exists x)\ \neg(\exists y)\ Q(x, y)$$

$$(\exists x)\ (\forall y)\ \neg Q(x, y).$$

This example suggests that it is permissible to permute a negation and a quantifier by changing the type of quantifier, and indeed this is so.

Let $\mathcal{Q}_i$ be a quantifier, for $1 \leq i \leq n$. For each $\mathcal{Q}_i$, let $\mathcal{Q}_i^*$ be the complementary quantifier. That is, if $\mathcal{Q}_i = \forall$, then let $\mathcal{Q}_i^* = \exists$; if $\mathcal{Q}_i = \exists$, then let $\mathcal{Q}_i^* = \forall$. Then,

$$\neg(\mathcal{Q}_1 x_1)(...)(\mathcal{Q}_n x_n)\ P(\bar{x}) \equiv (\mathcal{Q}_1^* x_1)(...)(\mathcal{Q}_n^* x_n)\ \neg P(\bar{x}).$$

3.5 Proof Strategies

There are two elementary logical forms—universal statements and existence proofs—that occur so commonly in mathematical claims that they warrant some general discussion.

3.5.1 Universal Statements. A logical form you are likely to encounter very often is

$$(\forall x)\ [H(x) \Rightarrow P(x)], \tag{3.14}$$

where $H(x)$ and $P(x)$ are formulas in one variable. Statements in this form are called *universal statements*. The formulas H and P are used to characterize properties of mathematical objects so that the claims in this form may be thought of as stating:

> If a mathematical object has property H, then it has property P as well.

This is particularly useful if we know a great deal about mathematical objects that have property P. Because the statement we are endeavoring to prove is universal, examples do not suffice to prove such claims—the example you cite might accidentally have properties H and P. Rather, universal claims must be proved abstractly, arguing that satisfying a definition or set of properties implies the satisfaction of other properties. This generally requires carefully evaluating definitions. In practice, we often do this by assuming that we have an arbitrary element that satisfies a definition, or explicit assumptions, and logically derive additional conclusions about this object. By "arbitrary" we mean that we are not allowed to make any claims about the element except those that follow immediately from definitions, explicit assumptions, or are logically derived from definitions and explicit assumptions. Since the object was arbitrary (except for the explicit assumptions you make at the outset of the argument), the conclusions you derive concerning the object will be true universally of all objects that satisfy the assumptions.

EXAMPLE 3.15. Suppose $F(x)$ is the formula:

$$x \in \mathbb{N} \text{ and } x \text{ is a multiple of } 4.$$

Let $E(x)$ be the formula:

x is even.

Then

$$(\forall x)\ [F(x) \Rightarrow E(x)]. \tag{3.16}$$

It does not suffice to observe that 4, 8, and 12 are all even. To argue for the statement directly, you would argue abstractly that any object that satisfies $F(x)$ necessarily satisfies $E(x)$.

There are a couple of approaches that one commonly considers when proving conditional statements. Choosing an approach is choosing a strategy for the proof. Normally, more than one strategy can be made to work, but often one may be simpler than the others.

Claims of the form (3.14) are generally approached in one of the following ways:

(1) Direct Proof

Let x be an object for which H holds. By decoding the property H, you might be able to show directly that P holds for x as well. Since x was an arbitrary object satisfying P, the universal claim will be proved.

EXAMPLE 3.17. Prove statement (3.16) directly.

Let $x \in \mathbb{N}$ (we treat x as a fixed but arbitrary element of the natural numbers). If $x = 4n$, then

$$x = 2 \cdot (2n)$$

and is therefore even.

EXAMPLE 3.18. Prove that any three points in the plane are either collinear or lie on a circle.

PROOF. Label the points A, B, C. Let L be the perpendicular bisector of AB. Every point on L is equidistant from A and B.

Let M be the perpendicular bisector of BC. Every point on M is equidistant from B and C.

If A, B, and C are not collinear, the lines L and M are not parallel, so they intersect at some point D. The point D is equidistant from A, B, and C, so these points lie on a circle centered at D. $\qquad\qquad$ $\square$

EXAMPLE 3.19. Pythagoras's theorem can be stated in the form (3.14). (What are H and P in this case?) Euclid's proof of Pythagoras's theorem is a direct proof (Euclid's *Elements* I.47).

(2) Contrapositive Proof

It is sometimes easier to show that the failure of P implies the failure of H. Assume you have an object for which P fails (that is, assume $\neg P$ holds for the object). Derive that H must fail for the object as well. In this case you will have demonstrated that

$$(\forall x) \, [\neg P(x) \Rightarrow \neg H(x)].$$

This is equivalent to the claim

$$(\forall x) \, [H(x) \Rightarrow P(x)].$$

EXAMPLE 3.20. Prove statement (3.16) by proving the contrapositive.

Let $x \in \mathbb{N}$, and assume $\neg E(x)$, so x is odd. As x is odd, then x divided by 4 has remainder 1 or 3. Then,

$$x \neq 4n.$$

So x is not a multiple of 4.

EXAMPLE 3.21. Prove that if x is an integer and x^2 is even, then x is even.

The contrapositive is the assertion that if x is an odd integer, then x^2 is odd. We shall prove this.

Suppose x is odd, so $x = 2n + 1$ for some integer n. Then $x^2 = 4n^2 + 4n + 1$, so $x^2 \equiv 1 \mod 2$, and x^2 is therefore odd.

(3) Contradiction

This is a proof in which we show that $H \wedge \neg P$ is necessarily false. That is, assume that H holds for an arbitrary object and P fails for that object, and show that this gives rise to a contradiction. Since contradictions are logically impossible, it is logically necessary that

$$\neg(H \wedge \neg P),$$

which is propositionally equivalent to

$$\neg H \vee P$$

or, alternatively,

$$H \Rightarrow P.$$

Since we shall have shown that for any substitution of x, the statement $H \Rightarrow P$ holds, we shall have shown the universal claim.

EXAMPLE 3.22. Prove statement (3.16) by contradiction.

Assume that x is a multiple of 4 and that x is odd. Let r be the residue of x modulo 2. Since x is a multiple of $4 = 2 \cdot 2$, we have that $r \equiv 0 \mod 2$. Since r is odd, we have that $r \equiv 1 \mod 2$. This implies $0 \equiv 1 \mod 2$, a contradiction. Therefore the assumption that there was an x that was both a multiple of 4 and odd is false, and so (3.16) must be true.

EXAMPLE 3.23. Prove that $\sqrt{2}$ is irrational.

PROOF. We restate this as an implication: If a number is rational, its square cannot equal 2. We begin by considering the logical structure of the claim. Here the hypothesis $H(x)$ is that x is a rational number, and the conclusion $P(x)$ is that $x^2 \neq 2$. We wish to prove

$$(\forall x)\ H(x) \Rightarrow P(x).$$

We shall give a proof by contradiction. That is, we assume the statement is false and derive a contradiction. So we assume

$$\neg((\forall x)\ H(x) \Rightarrow P(x)).$$

This is logically equivalent to

$$(\exists x)H(x) \wedge \neg(P(x)).$$

Let us go back to mathematical prose now that we have fought through the logic. Assume that x is a rational number, and assume also that $x^2 = 2$; we wish to derive a logical contradiction. Write $x = m/n$, where m and n are nonzero integers that have no common factors. Then

$$x^2 = m^2/n^2 = 2,$$

so $m^2 = 2n^2$. Therefore m^2 is even, so, by Example 3.21, m is even. Therefore $m = 2k$ for some integer k, and so

$$m^2 = 4k^2 = 2n^2.$$

Therefore $n^2 = 2k^2$ is even, so n is even. But then both m and n are even and so have 2 as a common factor, which contradicts the assumption that m/n was the reduced form of the rational number x. $\square$

Contrapositive proofs and proofs by contradiction are very similar. Indeed, any contrapositive proof, that $\neg P \Rightarrow \neg H$, automatically yields that $(H \wedge \neg P)$ is impossible. The distinction is more linguistic than logical. The reason for having names for different proof strategies is to provide guidance to the reader in order to make the proof easier to follow.

In Chapter 4 we shall see another powerful method for proving universal statements over $\mathbb{N}$, namely the Principle of Induction.

3.5.2 Existence Proofs. A second common form for a mathematical claim is an existential statement, that is, a statement in the form

$$(\exists x)\ P(x). \tag{3.24}$$

There are three common approaches to proving existential statements.

(1) Construction

Obviously, the most direct way to show that something exists with certain properties is to introduce or construct an object with property P. For claims in this form, the example is the proof, although you will need to show that the object satisfies P, if it is not obvious.

EXAMPLE 3.25. Prove that there exists a real function whose first derivative is everywhere positive and whose second derivative is everywhere negative.

PROOF. The easiest way to do this is to write down a function with these properties. One such function is $f(x) = 1 - e^{-x}$. The derivative is e^{-x}, which is everywhere positive, and the second derivative is $-e^{-x}$, which is everywhere negative.   □

(2) Counting

Sometimes one can establish an object's existence by a counting argument.

EXAMPLE 3.26. Suppose there are 30 students in a class. Show that at least two of them share the same last initial.

PROOF. For each letter A,B,... group all the students with that letter as their last initial. As there are only 26 groups and $30 > 26$ students, at least one group must have more than on student in it.   □
□

The argument we just gave is called the *pigeonhole principle* based on the analogy of putting letters into pigeonholes. If there are more letters than pigeonholes, then some pigeonhole must have more than one letter. Notice that unlike a constructive proof, a counting proof does not tell you which group has more than one element in it.

For Cantor's spectacular generalization of the pigeonhole principle to infinite sets, see Chapter 6.

(3) Contradiction

It can be difficult to prove existential statements by construction. An alternative is to assume that the existential statement is false (that there is no object which satisfies $P(x)$). If it is impossible that no object has property P, then some object must. Again, this approach may not give us much insight into the objects that have property P. See, for example, Exercise 3.27.

EXAMPLE 3.27. Suppose all the points in the plane are colored either red or blue. Prove that there must be two points of the same color exactly one unit apart.

PROOF. Assume there are not two such points. Draw an equilateral triangle of side 1. Label its vertices A, B, and C. Then A and B must be different colors, B and C must be different colors, and C and A must be different colors. This is impossible with only two colors to choose from. □

Notice that we have not said whether there is a red-red pair that is unit distance apart, or a blue-blue pair that is unit distance apart, just that one such pair must exist.

3.6 Exercises

EXERCISE 3.1. Prove de Morgan's laws, (3.3) and (3.4). (Hint: There are four possible assignments of truth values 0 and 1 to the two statements P and Q. For each such assignment, evaluate the truth values of the left-hand and right-hand sides of (3.3) and show they are always the same.)

EXERCISE 3.2. Prove that compound statements P and Q are propositionally equivalent iff $P \iff Q$.

EXERCISE 3.3. Give an example of a true conditional statement in which the consequence is false.

EXERCISE 3.4. If P, Q, and R are statements, prove that the following are true:

(i) $P \wedge \neg P \Rightarrow Q$
(ii) $[(P \Rightarrow Q) \wedge (Q \Rightarrow R)] \Rightarrow (P \Rightarrow R)$
(iii) $[P \Rightarrow (Q \wedge \neg Q)] \Rightarrow \neg P$
(iv) $[P \wedge (P \Rightarrow Q)] \Rightarrow Q$
(v) $P \Rightarrow (Q \vee \neg Q)$

EXERCISE 3.5. Let P and Q be statements. Prove that there are statements using only P, Q, $\neg$, and $\wedge$ that are propositionally equivalent to

(i) $P \wedge Q$
(ii) $P \vee Q$
(iii) $P \Rightarrow Q$

Prove that there are statements using only P, Q, $\neg$, and $\vee$ that are equivalent to the above.

EXERCISE 3.6. Prove the distributive laws for propositional logic: If P, Q and R are statements, then

(i) $P \vee (Q \wedge R) \equiv (P \vee Q) \wedge (P \vee R)$
(ii) $P \wedge (Q \vee R) \equiv (P \wedge Q) \vee (P \wedge R)$

EXERCISE 3.7. Prove the distributive law for sets: If X, Y and Z are sets, then

  (i) $X \cup (Y \cap Z) = (X \cup Y) \cap (X \cup Z)$

  (ii) $X \cap (Y \cup Z) = (X \cap Y) \cup (X \cap Z)$

EXERCISE 3.8. Let sets X, Y, and Z be characteristic sets of formulas $P(x)$, $Q(x)$, and $R(x)$ respectively. For each possible region of the Venn diagram of X, Y, and Z give a compound formula (with atomic formulas P, Q, and R) that has that region as its characteristic set.

EXERCISE 3.9. Write a formula in one variable that defines the even integers.

EXERCISE 3.10. Write a formula that defines perfect squares.

EXERCISE 3.11. Write a formula in two variables that defines the points in $\mathbb{R}^2$ that have distance 1 from the point (π, e).

EXERCISE 3.12. Can you write a formula in one variable using only addition, multiplication, exponentiation, integers, and equality to define the set of all roots of a given polynomial with integer coefficients? How about the set of roots of all polynomials with integer coefficients?

EXERCISE 3.13. Which of the following statements are true?

  (i) $(\forall x \in \mathbb{R})\; x + 1 > x$

  (ii) $(\forall x \in \mathbb{Z})\; x^2 > x$

  (iii) $(\exists x \in \mathbb{Z})(\forall y \in \mathbb{Z})\; x \leq y$

  (iv) $(\forall y \in \mathbb{Z})(\exists x \in \mathbb{Z})\; x \leq y$

  (v) $(\forall \varepsilon > 0)(\exists \delta > 0)(\forall x \in \mathbb{R})\; [0 < |\, x - 1\, | < \delta] \Rightarrow [|\, x^2 - 1\, | < \varepsilon]$

EXERCISE 3.14. What is the negation of each statement in Exercise 3.13? Which of the negations are true?

EXERCISE 3.15. Let a, $L \in \mathbb{R}$, and f be a real function. Prove that the statements

$$(\forall \varepsilon > 0)(\exists \delta > 0)(\forall x \in \mathrm{Dom}(f))\; [0 < |\, x - a\, | < \delta] \Rightarrow [|\, f(x) - L\, | < \varepsilon]$$

and

$$(\exists \delta > 0)(\forall \varepsilon > 0)(\forall x \in \mathrm{Dom}(f)) \; [0 < \mid x - a \mid < \delta] \Rightarrow [\mid f(x) - L \mid < \varepsilon]$$

are not equivalent. Which statement is a consequence of the other?

EXERCISE 3.16. Let $P(x, y)$ be a formula in two variables. Show that in general $(\forall x)(\exists y) \, P(x, y)$ need not be equivalent to $(\exists y)(\forall x) \, P(x, y)$. Show that $(\forall x)(\forall y) \, P(x, y)$ is equivalent to $(\forall y)(\forall x) \, P(x, y)$. What about $(\exists x)(\exists y) \, P(x, y)$ and $(\exists y)(\exists x) \, P(x, y)$?

EXERCISE 3.17. Consider the following statements. Write down the contrapositive and the converse to each one.

(i) All men are mortal.

(ii) I mean what I say.

(iii) Every continuous function on the interval $[0, 1]$ attains its maximum.

(iv) The sum of the angles of a triangle is 180°.

EXERCISE 3.18. Prove that a number is divisible by 4 iff (that is, if and only if) its last two digits are.

EXERCISE 3.19. Prove that a number is divisible by 8 iff its last three digits are.

EXERCISE 3.20. Prove that a number is divisible by 2^n iff its last n digits are.

EXERCISE 3.21. Suppose m is a number with the property that any natural number is divisible by m iff its last three digits are. What does this say about m? Prove your assertion.

EXERCISE 3.22. Prove that an integer is divisible by 11 iff the sum of the oddly placed digits minus the sum of the evenly placed digits is divisible by 11. (So $11 \mid 823493$ iff 11 divides $(2 + 4 + 3) - (8 + 3 + 9)$.)

EXERCISE 3.23. Show that every interval contains rational and irrational numbers.

EXERCISE 3.24. Prove that $\sqrt{3}$ is irrational.

EXERCISE 3.25. Prove that $\sqrt{10}$ is irrational.

EXERCISE 3.26. Prove that the square root of any natural number is either an integer or irrational.

EXERCISE 3.27. Prove that there exist irrational numbers x and y so that x^y is rational. (Hint: consider $\sqrt{2}^{\sqrt{2}}$ and $\left(\sqrt{2}^{\sqrt{2}}\right)^{\sqrt{2}}$.)

EXERCISE 3.28. Prove or disprove the following assertion: Any four points in the plane, no three of which are collinear, lie on a circle.

EXERCISE 3.29. Prove that there are an infinite number of primes.

EXERCISE 3.30. For $k = 0, 1, 2$, let P_k be the set of prime numbers that are congruent to $k \bmod 3$. By Exercise 3.29, $P_0 \cup P_1 \cup P_2$ is infinite. Can you say which of the sets P_0, P_1, and P_2 are infinite?

REMARK. For two of the three sets, this problem is not too difficult. For the third one, it is extremely difficult and is a special case of a celebrated theorem of Dirichlet. See, for example, [9] for a treatment of Dirichlet's theorem.

EXERCISE 3.31. Let the points in $\mathbb{R}^2$ be colored red, green, and blue. Prove that either there are two points of the same color a distance 1 apart, or there is an equilateral triangle of side length $\sqrt{3}$ all of whose vertices are the same color.

EXERCISE 3.32. Prove that

$$e = \sum_{n=0}^{\infty} \frac{1}{n!}$$

is irrational. (Hint: Argue by contradiction. Assume $e = \frac{p}{q}$, and multiply both sides by $q!$. Rearrange the equation to get an integer equal to an infinite sum of rational numbers that converges to a number in $(0, 1)$.)

CHAPTER 4

Principle of Induction

4.1 Well-Orderings

In this chapter we discuss the principle of mathematical induction. Be aware that the word "induction" has a different meaning in mathematics than in the rest of science. The principle of mathematical induction depends on the order structure of the natural numbers and gives us a powerful technique for proving universal mathematical claims.

DEFINITION. Well-ordering Let X be a set, and $\preceq$ a linear ordering on X. We say that X is well-ordered with respect to $\preceq$ (or $\preceq$ is a well-ordering of X) if every nonempty subset of X has a least element with respect to $\preceq$. That is, for any nonempty subset Y of X

$$(\exists a \in Y)(\forall y \in Y)\ a \preceq y.$$

In general, linear orderings need not be well-orderings. Well-ordering is a universal property—a set X with an ordering $\preceq$ is well-ordered if *every* nonempty subset of X has a least element with respect to $\preceq$. If there is any nonempty subset that does not have a least element, then $\preceq$ does not well-order X.

EXAMPLE 4.1. $\mathbb{Z}$ is not well-ordered by $\leq$. The integers do not have a least element, which suffices to demonstrate that $\mathbb{Z}$ is not well-ordered by $\leq$.

EXAMPLE 4.2. Let $X = \{x \in \mathbb{R} \mid x \geq 2\}$. Let $\leq$ be the usual ordering on $\mathbb{R}$. X is linearly ordered by $\leq$, but X is not well-ordered by $\leq$. In this example, X has a least element, but any open interval contained in X will fail to have a least element.

The key order properties of $\mathbb{N}$ are that it is well-ordered and every element of $\mathbb{N}$, except 0, is the successor of a natural number:

WELL-ORDERING PRINCIPLE FOR THE NATURAL NUMBERS: *The set $\mathbb{N}$ is well-ordered by $\leq$.*

SUCCESSOR PRINCIPLE FOR THE NATURAL NUMBERS: *If $n \in \mathbb{N}$ and $n \neq 0$, then there is $m \in \mathbb{N}$ such that $n = m + 1$.*

If one accepts an intuitive understanding of the natural numbers, these principles are more or less obvious. Indeed, let Y be any nonempty subset of $\mathbb{N}$. Since it is nonempty, there is some m in Y. Now, consider each of the finitely many numbers $0, 1, 2, \ldots, m$ in turn. If $0 \in Y$, then 0 is the least element. If 0 is not in Y, proceed to 1. If this is in Y, it must be the least element; otherwise proceed to 2. Continue in this way, and you will find some number less than or equal to m that is the least element of Y.

This argument, though convincing, does rely on the fact that we have an idea of what $\mathbb{N}$ "is." If we wish to define $\mathbb{N}$ in terms of set operations, as we do in Chapter 8, we essentially have to include the well-ordering principle for the natural numbers as an axiom.

4.2 Principle of Induction

We begin by proving a theorem that is equivalent to the principle of induction.

THEOREM 4.3. *If*

(1) $X \subseteq \mathbb{N}$

(2) $0 \in X$

(3) $(\forall n \in \mathbb{N})\ n \in X \Rightarrow (n + 1) \in X$

then

$$X = \mathbb{N}.$$

DISCUSSION. *We shall argue by contradiction. We assume that $X \neq \mathbb{N}$. Let Y be the complement of X in $\mathbb{N}$. Since Y is nonempty, it will have a least element. The third hypothesis of the theorem will not permit a least element in Y, other than 0, and this is impossible by the second hypothesis. Therefore Y is necessarily empty.*

PROOF. Let X satisfy the hypotheses of the theorem. Let

$$Y = \mathbb{N} \setminus X.$$

We assume Y is nonempty. Since $Y \subseteq \mathbb{N}$, Y is well-ordered by $\leq$. Let $a \in Y$ be the least element of Y. We note that a is not 0, since $0 \in X$. Therefore $a \geq 1$ and is a successor, so $a - 1$ is in $\mathbb{N}$ and not in Y. Hence $a - 1$ is in X. But then by hypothesis (3) of the theorem, $a - 1 + 1 \in X$. This is a contradiction; therefore, Y is empty and $X = \mathbb{N}$. ☐

REMARK. We will occasionally include informal, labelled discussions in our proofs to guide you in your reading. This is not a usual practice. You should not include such discussions in your proofs unless your instructor requests it.

Theorem 4.3 is more easily applied in the following form.

COROLLARY 4.4. *Principle of induction* *Let $P(x)$ be a formula in one variable. If*

(1) $P(0)$

(2) $(\forall x \in \mathbb{N}) \ P(x) \Rightarrow P(x + 1)$

then

$$(\forall x \in \mathbb{N}) \ P(x).$$

PROOF. Let

$$\chi_P \; = \; \{x \in \mathbb{N} \mid P(x)\}.$$

We wish to show that $\chi_P \; = \; \mathbb{N}$. By assumption (1) $P(0)$, so $0 \in \chi_P$. Assume that $n \in \chi_P$. Then $P(n)$. By assumption (2)

$$P(n) \Rightarrow P(n+1).$$

Therefore $P(n+1)$ and $n+1 \in \chi_P$. Since n is arbitrary,

$$(\forall n \in \mathbb{N}) \; n \in \chi_P \Rightarrow n+1 \in \chi_P.$$

By Theorem 4.3, $\chi_P \; = \; \mathbb{N}$ and

$$(\forall x \in \mathbb{N}) \; P(x).$$

$\square$

Suppose that you wish to show that a formula $P(x)$ holds for all natural numbers. When arguing by induction, the author must show that the hypotheses for the theorem are satisfied. Typically, the author first proves that $P(0)$. This is called the   base case   of the proof by induction. It is very often an easy, even trivial, conclusion. Nonetheless, it is necessary to prove a base case in order to argue by induction (can you demonstrate this?). Having proved the base case, the author will then prove the second hypothesis, namely, that the claim being true for an arbitrary natural number implies that it is true at the successor of that natural number. This is the   induction step . The induction step requires proving a conditional statement, which is often proved directly. It is important to understand that the author is not claiming that P holds at an arbitrary natural number; otherwise the argument would be circular and invalid. Rather, the author will demonstrate that if the result *were* true at an arbitrary natural number, then it *would be* true for the subsequent natural number. The assumption that P holds at a fixed and arbitrary natural number is called the   induction hypothesis . If the author successfully proves the base case and the induction step, then the assumptions of Corollary 4.4 are satisfied, and P holds at all natural numbers.

PROPOSITION 4.5. *Let $N \in \mathbb{N}$. Then*

$$\sum_{n=0}^{N} n = \frac{N(N+1)}{2}.$$

DISCUSSION. *This is a good first example of a proof by induction. The argument is a straightforward application of the technique and the result is of historical and practical interest.*

We argue by induction on the upper index of the sum. That is, the formula we are proving for all natural numbers is

$$P(x): \quad \sum_{n=0}^{x} n = \frac{x(x+1)}{2}.$$

It is important to identify the quantity over which you are applying the principle of induction, but some authors who are writing an argument for readers who are familiar with induction may not explicitly state the formula.

We prove a base case, $N = 0$, that corresponds to the sum with the single term 0. We then argue the induction step. This is our first argument using the principle of induction. Pay close attention to the structure of this proof. You should strive to follow the conventions for proofs by induction that we establish in this book.

PROOF. BASE CASE: $N = 0$.

DISCUSSION. *Note that the base case is the statement $P(0)$.*

Since

$$\sum_{n=0}^{0} n = 0 = \frac{(0)(1)}{2},$$

$P(0)$ holds.

INDUCTION STEP:

DISCUSSION. *We prove the universal statement*

$$(\forall\, x \in \mathbb{N})\ P(x) \Rightarrow P(x+1).$$

by showing that for an arbitrary natural number N,

$$P(N) \Rightarrow P(N+1).$$

Thus we reduce proving a universal statement to proving an abstract conditional statement. We prove the resulting conditional statement directly. That is, we assume $P(N)$ *and derive* $P(N+1)$. *We remind the reader that we are not claiming the result holds at* N—*that is, we do not claim* $P(N)$. *Rather, we are proving the conditional statement by assuming the antecedent (the induction hypothesis) and deriving the consequence. If you do not use the induction hypothesis, you are not arguing by induction. Of course, in the body of the argument this is transparent without reference to the underlying logical principles.*

Let $N \in \mathbb{N}$ and assume that

$$\sum_{n=0}^{N} n = \frac{N(N+1)}{2}.$$

Then

$$\sum_{n=0}^{N+1} n = \left(\sum_{n=0}^{N} n\right) + N + 1$$

$$=_{IH} \frac{N(N+1)}{2} + N + 1$$

by the induction hypothesis.

DISCUSSION. *It is a good habit, and a consideration for your reader, to identify when you are invoking the induction hypothesis. We will use the subscript $_{IH}$ to indicate where we invoke the induction hypothesis.*

So

$$\sum_{n\,=\,0}^{N+1} n \;=\; \frac{N(N+1)}{2} + N + 1$$

$$=\; \frac{N(N+1)}{2} + \frac{2N+2}{2}$$

$$=\; \frac{N^2 + 3N + 2}{2}$$

$$=\; \frac{(N+1)((N+1)+1)}{2}.$$

Therefore,

$$(\forall\, N \in \mathbb{N})\ P(N) \Rightarrow P(N+1).$$

By the principle of induction, the proposition follows.   □

PROPOSITION 4.6. *Let $N \in \mathbb{N}$. Then*

$$\sum_{n\,=\,0}^{N} n^2 \;=\; \frac{N(N+1)(2N+1)}{6}. \tag{4.7}$$

PROOF. The assertion $P(N)$ is that the equation (4.7) holds. The base case, $N = 0$, is obvious:

$$\sum_{n\,=\,0}^{0} n^2 \;=\; \frac{0(0+1)(2 \cdot 0 + 1)}{6}.$$

INDUCTION STEP:
Assume that $N \in \mathbb{N}$ and

$$\sum_{n\,=\,0}^{N} n^2 \;=\; \frac{N(N+1)(2N+1)}{6}.$$

We prove that

$$\sum_{n=0}^{N+1} n^2 = \frac{(N+1)(N+2)(2N+3)}{6}.$$

Indeed

$$
\begin{aligned}
\sum_{n=0}^{N+1} n^2 &= \left(\sum_{n=0}^{N} n^2 \right) + (N+1)^2 \\
&\underset{IH}{=} \frac{N(N+1)(2N+1)}{6} + (N+1)^2. \\
&= \frac{N(N+1)(2N+1)}{6} + (N+1)^2 \\
&= \frac{2N^3 + 9N^2 + 13N + 6}{6} \\
&= \frac{(N+1)(N+2)(2(N+1)+1)}{6}.
\end{aligned}
$$

The proposition follows from the principle of induction. $\qquad\qquad$ $\square$

DISCUSSION. *The proof of Proposition 4.6 is very similar to the proof of Proposition 4.5. You may wish to confirm the algebraic identities in the latter portion of the proof since they are not obvious. Just enough detail is included to guide you through the proof of the implication. The author of a proof by induction will assume that you are comfortable with the technique and thereby may provide less detail than you like.*

REMARK. There is more to Propositions 4.5 and 4.6 than just the proofs. There are also the formulas. Indeed, one use of induction is that if you *guess* a formula, you can use induction to prove your formula is correct. See Exercises 4.12 and 4.16.

Why is a base case necessary? Consider the following argument for the false claim

$$\sum_{n=0}^{N} n < \frac{N(N+1)}{2}.$$

Let $N \in \mathbb{N}$ and assume $P(N)$, where $P(N)$ is the statement

$$\sum_{n=0}^{N} n < \frac{N(N+1)}{2}.$$

Then

$$\sum_{n=0}^{N+1} n = \left(\sum_{n=0}^{N} n \right) + N + 1$$

$$<_{IH} \frac{N(N+1)}{2} + N + 1$$

$$= \frac{N^2 + 3N + 2}{2}$$

$$= \frac{(N+1)((N+1)+1)}{2}.$$

Hence,

$$(\forall N \in \mathbb{N})\ P(N) \Rightarrow P(N+1).$$

Of course the inequality $P(N)$ is easily demonstrated to be false. What went wrong? Without a base case, proving

$$(\forall N \in \mathbb{N})\ P(N) \Rightarrow P(N+1)$$

is not sufficient to prove $(\forall N \in \mathbb{N})\ P(N)$. If $P(0)$ were true, then $P(1)$ would be true, and if $P(1)$ were true, then $P(2)$ would be, and so on. Indeed, if we are able to prove $P(N)$ for any $N \in \mathbb{N}$, then we know $P(M)$ for any natural number $M > N$. But the sequence of statements $\langle P(0), P(1), P(2), \ldots \rangle$ never gets started. $P(N)$ fails for all N.

Another way to think of induction is in terms of guarantees. Suppose you decide to buy a car. First you go to Honest Bob's. Bob guarantees that any car he sells will go at least one mile. You buy a car, drive it off the lot, and after three miles it breaks down and cannot be fixed. You walk back angrily, but Bob will not give you your money back because the car lived up to the guarantee.

Then you cross the road to Honest John's. John guarantees that if he sells you a car, once it starts it will never stop. This sounds pretty good, so you buy a car, put the keys in the ignition, and ... nothing. The car will not start. John will not give you your money back either because the car did not fail to do what he claimed.

Feeling desperate, you end up at Honest Leo's. Leo's cars come with two guarantees:

(1) The car will start and go at least one mile.
(2) No matter how far the car has gone, it can always be driven an extra mile.

You think this over, and eventually decide that the car will go forever. Best of all, the lease is only $1 a month for the first two months. You sign the lease and drive home rather pleased with yourself.[1]

There are many handy generalizations of the principle of induction. The first we discuss is called strong induction. It is so named because the induction hypothesis is stronger than the induction hypothesis in standard induction, and hence the induction step is sometimes easier to prove in an argument by strong induction.

COROLLARY 4.8. *Strong induction* *Let $P(x)$ be a formula such that*

(1) $P(0)$

---

[1] You are correct that the principle of induction guarantees that your car will drive forever. However, as your mother points out when you show her the lease, after the first two months your payment each month is the sum of your payments in the previous two months. How much will you be paying after five years?

(2) *For each $n \in \mathbb{N}$,*

$$[(\forall\, x < n)\, P(x)] \Rightarrow [P(n)].$$

Then

$$(\forall\, x \in \mathbb{N})\ P(x).$$

Intuitively this is not very different from basic induction. You start at a base case, and once started you can continue through the remainder of the natural numbers. The distinction is just in the number of assumptions you use when proving something by strong induction. In practice, it gives the advantage that in the induction step you can reduce case N to any previous case rather than the immediately preceding case, $N - 1$. In particular this simplifies arguments about divisibility and integers.

DISCUSSION. *We reduce the principle of strong induction to the principle of induction. We accomplish this by introducing a formula, $Q(x)$, which says, "$P(y)$ is true for all $y < x$." Strong induction on $P(x)$ is equivalent to basic induction on $Q(x)$.*

PROOF. Assume that $P(x)$ satisfies the hypotheses of Corollary 4.8. Let $Q(x)$ be the formula

$$(\forall\, y \le x)\ P(y)$$

where the universe of y is $\mathbb{N}$. Then $Q(0) \equiv P(0)$, so is true. Let $N \in \mathbb{N}$, $N \ge 1$, and assume $Q(N)$. So

$$(\forall\, y \le N)\ P(y)$$

and therefore $P(N + 1)$. Hence

$$(\forall\, n \le N + 1)\ P(y)$$

and thus $Q(N + 1)$. Therefore

$$(\forall\, x \in \mathbb{N})\ Q(x) \Rightarrow Q(x + 1).$$

By the principle of induction,

$$(\forall\, x \in \mathbb{N})\ Q(x).$$

However, for any $N \in \mathbb{N}$, $Q(N) \Rightarrow P(N)$, so

$$(\forall\, x \in \mathbb{N})\ P(x).$$

$\square$

Strong induction is particularly useful when proving claims about division. There are examples of the technique throughout Chapter 7. The results in Chapter 7 do not require Chapter 5 and Chapter 6, so you may easily skip ahead. See, for example, Section 7.1, where the Fundamental Theorem of Arithemetic is proved using strong induction.

Induction does not have to start at 0 or even at a natural number.

COROLLARY 4.9. *Let $k \in \mathbb{Z}$, and $P(x)$ be a formula in one variable such that*

    (1) $P(k)$

    (2) $(\forall\, x \geq k)\ P(x) \Rightarrow P(x+1)$.

Then

$$(\forall\, x \in \mathbb{Z})\ x \geq k \Rightarrow P(x).$$

DISCUSSION. *This can be proved by a defining a new formula that can be proved with standard induction. Can you define the formula?*

4.3 Polynomials

We now use the machinery developed in Section 4.2 to undertake a modest mathematical program. As we indicated in Chapter 0, most of you, until now, have used mathematical results to solve problems in computation. Here we are interested in proving a result with which you may be familiar.

This result concerns polynomials with real coefficients (i.e., coefficients that are real numbers). You have spent a good deal of your mathematical life investigating polynomials and undoubtedly can make many interesting and truthful claims about them. But how confident are you that these claims are true? It is possible that your belief in these claims is, by and large, mere confidence in the claims and beliefs of experts in the field. In practice, you could do worse than to acquiesce to the assertions of specialists, and practical limitations generally compel us to accept many claims on faith. Of course, this practice carries risks. For hundreds of years, the assertions of Aristotle were broadly accepted, often in spite of empirical evidence to the contrary. Naturally, we continue to accept many claims on faith. In the case of modern science, we generally do not have firsthand access to primary evidence on which modern scientific theories are based. Mathematics is different from every other field of intellectual endeavor because you have the opportunity to verify virtually every mathematical claim you encounter. You are now at the point in your mathematical career at which you can directly confirm mathematical results.

The theorem we wish to prove is that the number of real roots of a real polynomial is, at most, the degree of the polynomial. You may be familiar with this claim but uncertain of why it holds. This result is interesting, in part, because it guarantees that the graph of a polynomial will cross any horizontal line only finitely many times. Put another way, level sets of polynomials cannot have more elements than the degree of the polynomial.

NOTATION. $\mathbb{R}[x]$ $\mathbb{R}[x]$ is the set of polynomials with real coefficients in the variable x.

THEOREM 4.10. *Let $N \in \mathbb{N}$ and $p \in \mathbb{R}[x]$ have degree $N \geq 1$. Then p has at most N real roots.*

DISCUSSION. *This result is sufficiently difficult that we shall have to prove three preliminary results. These lemmas[2] are proved within the argument for the theorem. Throughout the argument we shall be investigating a general polynomial, p, of degree N.*

PROOF. We prove first that the distributive property generalizes to an arbitrary number of summands.

LEMMA 4.11. *Let $N \in \mathbb{N}^+$ and, for $0 \leq n \leq N$, $a_n \in \mathbb{R}$. If $c \in \mathbb{R}$, then*

$$\sum_{n=0}^{N} ca_n = c \left(\sum_{n=0}^{N} a_n \right).$$

DISCUSSION. *This result generalizes the distributive property to more than two summands. We are assuming the distributive property of real numbers: for $a, b, c \in \mathbb{R}$,*

$$c \cdot (a + b) = ca + cb.$$

We prove the lemma by induction. It is surprising that a claim that seems so obvious uses the powerful machinery of induction. But remember that we are proving this for all finite sums of arbitrarily many summands. Of course, you may feel that the lemma is altogether obvious. If so, you should try to produce your own proof, or read this one for practice in mathematical induction in a context where the mathematical content is easy.

We shall argue by induction on the number of terms in the sum. The base case is for sums with two summands—this is just the distributive property. In the induction step we prove the conditional result that if the lemma holds for all sums with N terms, then it holds for all sums with $N + 1$ terms. At each step of the argument (base and induction

---

[2]A *lemma* is an auxiliary result that one uses in the proof of a theorem—sort of like a subroutine. In German, a theorem is called *Satz* and a lemma is called *Hilfsatz*, a "helper theorem."

steps) we are arguing for infinitely many concrete claims by arguing for a single abstract claim.

PROOF. We argue by induction on N.

BASE CASE: $N = 1$

Let $c, a_0, a_1 \in \mathbb{R}$. By the distributive property,

$$
\begin{aligned}
\sum_{n=0}^{1} ca_n &= ca_0 + ca_1 \\
&= c(a_0 + a_1) \\
&= c\left(\sum_{n=0}^{1} a_n\right).
\end{aligned}
$$

INDUCTION STEP:

Let $c \in \mathbb{R}$ and $a_n \in \mathbb{R}$, for $0 \leq n \leq N+1$. We assume

$$
\sum_{n=0}^{N} ca_n = c\left(\sum_{n=0}^{N} a_n\right).
$$

We have

$$
\begin{aligned}
\sum_{n=0}^{N+1} ca_n &= \left(\sum_{n=0}^{N} ca_n\right) + ca_{N+1} \\
&=_{IH} c\left(\sum_{n=0}^{N} a_n\right) + ca_{N+1}.
\end{aligned}
$$

By the distributive law (for two summands)

$$
\begin{aligned}
c\left(\sum_{n=0}^{N} a_n\right) + ca_{N+1} &= c\left(\sum_{n=0}^{N} a_n + a_{N+1}\right) \\
&= c\left(\sum_{n=0}^{N+1} a_n\right).
\end{aligned}
$$

Therefore,

$$\sum_{n=0}^{N+1} ca_n = c\left(\sum_{n=0}^{N+1} a_n\right).$$

By the induction principle, the result holds for all $N \in \mathbb{N}$. $\square$

LEMMA 4.12. *If $x, y \in \mathbb{R}$ and $n \in \mathbb{N}^+$, then*

$$x^n - y^n = (x-y)(x^{n-1} + x^{n-2}y + \cdots + xy^{n-2} + y^{n-1})$$

$$= (x-y)\left(\sum_{\substack{i,j\in\mathbb{N} \\ i+j = n-1}} x^i y^j\right).$$

DISCUSSION. *The notation in the last line of the lemma means that the sum is taken over all natural numbers i and j that have the property that $i + j = n - 1$.*

PROOF. By Lemma 4.11,

$$(x-y)\left(\sum_{\substack{i,j\in\mathbb{N} \\ i+j = n-1}} x^i y^j\right) = x\left(\sum_{\substack{i,j\in\mathbb{N} \\ i+j = n-1}} x^i y^j\right) - y\left(\sum_{\substack{i,j\in\mathbb{N} \\ i+j = n-1}} x^i y^j\right)$$

$$= \sum_{\substack{i,j\in\mathbb{N} \\ i+j = n-1}} x^{i+1} y^j - \sum_{\substack{i,j\in\mathbb{N} \\ i+j = n-1}} x^i y^{j+1}$$

$$= x^n - y^n.$$ $\square$

The next lemma associates roots of polynomials and linear factors.

LEMMA 4.13. *Let p be a polynomial of degree N. A real number, c, is a root of p iff*

$$p(x) = (x-c)q(x),$$

where $q(x)$ is a polynomial of degree $N - 1$.

DISCUSSION. *This lemma is a biconditional statement. That is, the lemma is propositionally equivalent to the conjunction of two conditional statements. We prove the conditional statements independently. One of the conditional statements is obvious (can you determine which?). The more difficult conditional statement will use Lemma 4.12. When proving a biconditional, $P \iff Q$, by proving the conditional statements $P \Rightarrow Q$ and $Q \Rightarrow P$, we often use ($\Rightarrow$) and ($\Leftarrow$) to identify the conditional statement under consideration.*

PROOF. Let p be a polynomial of degree N. Then there are a_0, a_1, ..., $a_N \in \mathbb{R}$, $a_N \neq 0$, such that,

$$p(x) = \sum_{n=0}^{N} a_n x^n.$$

($\Leftarrow$) Assume that there is $c \in \mathbb{R}$ and a polynomial q of degree $N - 1$ such that

$$p(x) = (x - c)q(x).$$

Then

$$p(c) = (c - c)q(c) = 0.$$

So c is a root of p.

($\Rightarrow$) Let $c \in \mathbb{R}$ be a root of p. Then

$$\begin{aligned} p(x) &= p(x) - p(c) \\ &= a_0 - a_0 + \sum_{n=1}^{N} a_n(x^n - c^n) \\ &= \sum_{n=1}^{N} a_n(x^n - c^n). \end{aligned}$$

By Lemma 4.12, for $n \geq 1$,

$$x^n - c^n = (x - c)q_n(x)$$

where

$$q_n(x) \; = \; x^{n-1} + cx^{n-2} + \cdots + c^{n-2}x + c^{n-1} \; = \; \sum_{\substack{i,j \in \mathbb{N} \\ i+j \,=\, n-1}} x^i c^j.$$

By Lemma 4.11,

$$p(x) \; = \; \sum_{n=1}^{N} a_n(x^n - c^n) \; = \; (x - c) \sum_{n=1}^{N} a_n q_n(x).$$

Let

$$q(x) \; = \; \sum_{n=1}^{N} a_n q_n(x).$$

For all n between 1 and N, $q_n(x)$ has degree $(n-1)$. So the degree of $q(x)$ is less than N. However the coefficient of x^{N-1} in $q(x)$ is a_N, and $a_N \neq 0$ by assumption. So the degree of $q(x)$ is $N - 1$, and

$$p(x) \; = \; (x - c)q(x).$$

$\square$

We complete the proof of Theorem 4.10. Let p be a polynomial of degree N. We argue by induction on the degree of p.

BASE CASE: $N = 1$.

If p is a polynomial of degree 1, then it is of the form

$$p(x) \; = \; a_1 x + a_0,$$

and the only root is $-a_0/a_1$.

INDUCTION STEP:

Assume that the theorem holds for $N \in \mathbb{N}^+$. Let p have degree $N + 1$. If p has no roots, the theorem holds for p. So assume that p has a real root, $c \in \mathbb{R}$. By Lemma 4.13,

$$p(x) \; = \; (x - c)q(x), \tag{4.14}$$

where q is of degree N. By the induction hypothesis, q has at most N real roots. If x is a root of p, then by (4.14) either x is a root of q or

$x = c$. Therefore p has at most $N + 1$ roots, proving the induction step. □

As a function, a polynomial in a particular variable is the same as a polynomial with the same coefficients in a different variable. Let $p \in \mathbb{R}[x]$ be

$$p(x) = \sum_{n=0}^{N} a_n x^n,$$

and $q \in \mathbb{R}[y]$ be

$$q(y) = \sum_{n=0}^{N} a_n y^n.$$

Then as real functions, p and q are the same function. That is,

$$\text{graph}(p) = \text{graph}(q).$$

As algebraic objects, however, one might occasionally wish to distinguish between polynomials in distinct variables.

We end this section by proving that polynomials are equal as functions if and only if they have the same coefficients.

COROLLARY 4.15. *Let* $p, q \in \mathbb{R}[x]$. *The coefficients of* p *and* q *are equal iff*

$$(\forall\, x \in \mathbb{R})\;\; p(x) = q(x).$$

PROOF. ($\Rightarrow$) If the coefficients of p and q are all equal, then, letting a_n denote the nth coefficient, we have

$$(\forall\, x \in \mathbb{R})\; p(x) = \sum_{n=0}^{N} a_n x^n = q(x).$$

($\Leftarrow$) Suppose $(\forall\, x \in \mathbb{R})\; p(x) = q(x)$. Then $p - q$ is a polynomial with infinitely many roots. If p and q disagree on any coefficient, then $p - q$ is a nonzero polynomial, has a degree, and, by Theorem 4.10, finitely many roots. Therefore, p and q must agree on all coefficients. □

4.4 Arithmetic-Geometric Inequality

We have presented modest generalizations of basic mathematical induction (Corollaries 4.8 and 4.9). The formality of our approach might suggest that induction is a rigid technique that must be applied inflexibly in a specific prescriptive way. To a mathematician, induction is governed by two ideas:

(1) Induction uses the well-ordering of the natural numbers, or more generally any well-ordered set, to prove universal statements quantified over the set.

(2) Every element in the set over which you quantify must be accounted for by the induction.

The formal characterizations of induction in Section 4.2 are sufficient but not necessary to achieve the objectives of a proof by induction. Theorem 4.16 will give you a sense about how the technique of induction can be extended.

DEFINITION.   Arithmetic mean   Let $a_1, \ldots, a_N$ be real numbers. The arithmetic mean of $a_1, \ldots, a_N$ is

$$\frac{1}{N}\left(\sum_{n=1}^{N} a_n\right).$$

DEFINITION.   Geometric mean   Let $a_1, \ldots, a_N$ be positive real numbers. The geometric mean of $a_1, \ldots, a_N$ is

$$\sqrt[N]{a_1 \cdots a_N}.$$

THEOREM 4.16.   *Arithmetic-geometric mean inequality*   *Let* $a_1, \ldots, a_N \in \mathbb{R}^+$. *Then*

$$\sqrt[N]{a_1 \cdots a_n} \leq \frac{1}{N}\left(\sum_{n=1}^{N} a_n\right). \tag{4.17}$$

DISCUSSION. *We prove this with an interesting argument due originally to Cauchy; our treatment is from the book* [1]. *We argue by* *induction on the size of the sample over which we are computing the means. After arguing the base case we show that if the inequality holds for the arithmetic and geometric means of N numbers, it necessarily holds for the means of $2N$ numbers. This implies that the theorem holds for the means of 2^N numbers for any $N \in \mathbb{N}$ (by a standard induction argument).*

We then show that the result holding for N numbers implies that it holds for $N - 1$ numbers. This implies that if the result holds at a natural number N, the inequality holds for all means of fewer than N numbers. Given any $k \in \mathbb{N}$, $2^k > k$, and since the theorem holds for means of 2^k numbers, it holds for means of k terms.

PROOF. We argue by induction on the number of terms on each side of the inequality.

BASE CASE: $(N = 2)$

Let $a_1, a_2 \in \mathbb{R}^+$. Then

$$(a_1 - a_2)^2 = a_1^2 - 2a_1a_2 + a_2^2 \geq 0.$$

Therefore

$$2a_1a_2 \leq a_1^2 + a_2^2$$

and

$$4a_1a_2 \leq a_1^2 + 2a_1a_2 + a_2^2$$
$$= (a_1 + a_2)^2.$$

Thus

$$2\sqrt{a_1a_2} \leq a_1 + a_2.$$

Therefore the inequality holds for two terms.

INDUCTION STEP:

Let $P(N)$ be the statement that (4.17) holds for all $a_1, \ldots, a_N > 0$. We show that $P(N) \Rightarrow P(2N)$. Let

$$G_N = \prod_{n=1}^{N} a_n$$

and

$$A_N = \left(\frac{\sum_{n=1}^{N} a_n}{N} \right).$$

So

$$
\begin{aligned}
G_{2N} &= \prod_{n=1}^{2N} a_n \\
&= \left(\prod_{n=1}^{N} a_n \right) \left(\prod_{n=N+1}^{2N} a_n \right) \\
&\underset{IH}{\leq} \left(\sum_{n=1}^{N} \frac{a_n}{N} \right)^N \left(\sum_{n=N+1}^{2N} \frac{a_n}{N} \right)^N.
\end{aligned}
$$

Let

$$B = \sum_{n=N+1}^{2N} \frac{a_n}{N}.$$

By the base case

$$
\begin{aligned}
A_N B &\leq \left(\frac{A_N + B}{2} \right)^2 \\
&= (A_{2N})^2.
\end{aligned}
$$

So

$$
\begin{aligned}
(A_N)^N B^N &= (A_N B)^N \\
&\leq ((A_{2N})^2)^N \\
&= (A_{2N})^{2N}.
\end{aligned}
$$

Thus

$$G_{2N} \leq (A_{2N})^{2N}.$$

Therefore, for any $N \in \mathbb{N}^+$,

$$P(N) \Rightarrow P(2N).$$

DISCUSSION. *Let $Q(N)$ be the statement $P(2^N)$. Then the argument thus far is a standard proof by induction of $(\forall N \in \mathbb{N}^+)\, Q(N)$. Of course we wish to show $(\forall N \in \mathbb{N})\, P(N)$. We do this by proving*

$$(\forall N \in \mathbb{N}^+)\, P(N+1) \Rightarrow P(N).$$

Let $N > 2$. We prove that

$$P(N+1) \Rightarrow P(N).$$

Assume $P(N+1)$. Then

$$(G_N)(A_N) \leq \left(\frac{(\sum_{n=1}^{N} a_n) + A_N}{N+1} \right)^{N+1}. \qquad (4.18)$$

DISCUSSION. *Recall that G_N is the product of $a_1, \ldots, a_N$. We are treating the sum A_N as the $N + 1^{st}$ factor, a_{N+1}, and applying the inequality $P(N+1)$.*

As

$$\left(\frac{(\sum_{n=1}^{N} a_n) + A_N}{N+1} \right)^{N+1} = \left(\frac{N A_N + A_N}{N+1} \right)^{N+1}$$

$$= (A_N)^{N+1},$$

inequality 4.18 gives

$$G_N A_N \leq A_N^{N+1},$$

and so

$$G_N \leq (A_N)^{N},$$

which is the statement $P(N)$. So

$$(\forall\, N \in \mathbb{N}^{+})\, P(N+1) \Rightarrow P(N).$$

Hence for all $N \geq 2$, $P(N)$.   □

The arithmetic mean and geometric mean are different ways of understanding averages. They are related by the arithmetic-geometric mean inequality (called the AGM inequality). Can we apply the inequality? Let us consider an easy geometrical application of the case $N = 2$. Consider the rectangle with sides of length a and b. The perimeter of the rectangle is

$$P \;=\; 2a + 2b,$$

and the area is

$$A \;=\; ab.$$

$$Perimeter = 2a{+}2b$$

$$a \qquad Area = ab$$

$$b$$

In calculus you proved that the rectangle of fixed perimeter with the greatest area is the square. This can also be proved directly from the AGM inequality:

$$
\begin{aligned}
P &= 2a + 2b \\
&= \frac{4a + 4b}{2} \\
&\geq \sqrt{16ab} \\
&= 4\sqrt{ab}.
\end{aligned}
$$

So

$$
\frac{P^2}{16} \geq ab = A.
$$

Recall that P is fixed and therefore so is $\frac{P^2}{16}$, and we have shown that this is an upper bound for the area of the rectangle.

Is this upper bound achieved? The area A of the rectangle varies according to the dimensions of the rectangle and if $a = b$

$$
\frac{P^2}{16} = \frac{(4a)^2}{16} = A.
$$

Thus the maximum area of the rectangle is achieved when $a = b$. This result can be generalized to higher dimensions—without the need for multivariable calculus.

Proving theorems is not just a question of technique, though this must be mastered. It also requires creativity and insight. A beautiful collection of proofs is contained in the book [1] by Martin Aigner and Günter Ziegler.

4.5 Exercises

EXERCISE 4.1. Prove by induction that 3 divides $7^n - 4$ for every $n \in \mathbb{N}^+$.

EXERCISE 4.2. Prove by induction that

$$
(\forall n \in \mathbb{N}) \ 2^n > n.
$$

EXERCISE 4.3. Prove that any subset of a well-ordered set is well-ordered.

EXERCISE 4.4. Prove that $(1+x)^n \geq 1 + nx$ for every $n \in \mathbb{N}^+$ and every $x \in (-1, \infty)$.

EXERCISE 4.5. Prove by induction that every finite set of real numbers has a largest element.

EXERCISE 4.6. Let X and Y be sets with n elements each. How many bijections from X to Y are there? What does this tell you about the number of permutations of $\ulcorner n \urcorner$? Prove your claim.

EXERCISE 4.7. The binomial coefficients $\binom{n}{k}$ can be defined from Pascal's triangle by

(i) $\forall n \in \mathbb{N}$, $\binom{n}{0} = \binom{n}{n} = 1$

(ii) $\forall 2 \leq n \in \mathbb{N}, \forall 1 \leq k \leq n-1$, $\binom{n}{k} = \binom{n-1}{k} + \binom{n-1}{k-1}$

Prove by induction that
$$\binom{n}{k} = \frac{n!}{(n-k)!k!}.$$

EXERCISE 4.8. Prove the binomial theorem: with $\binom{n}{k}$ defined by Exercise 4.7, for any $n \in \mathbb{N}$, the following identity holds
$$(x+y)^n = \sum_{k=0}^{n} \binom{n}{k} x^{n-k} y^k.$$

EXERCISE 4.9. Prove $\sum_{k=0}^{n} \binom{n}{k} = 2^n$.

EXERCISE 4.10. Prove, for all $n \in \mathbb{N}^+$, that
$$\binom{2n}{n} \geq \frac{2^{2n-1}}{\sqrt{n}}.$$

EXERCISE 4.11. The principle of descent says that there is no strictly decreasing, infinite sequence of natural numbers. Prove the principle of descent.

EXERCISE 4.12. The Fibonacci numbers are defined recursively by $F_1 = 1, F_2 = 1$, and for $n \geq 3$, $F_n = F_{n-1} + F_{n-2}$. Prove that the Fibonacci numbers are given by the equation

$$F_n = \frac{(1 + \sqrt{5})^n - (1 - \sqrt{5})^n}{2^n \sqrt{5}}. \tag{4.19}$$

This is an example of a formula that is hard to guess but easy to verify. For an explanation of how the formula arises, see Exercise 5.29.

EXERCISE 4.13. Let X be a set well-ordered by a relation $\preceq$. We say that a sequence of elements in X, $\langle x_n \mid n \in \mathbb{N} \rangle$, is strictly decreasing (with respect to $\preceq$) if for all $m, n \in \mathbb{N}$

$$[m < n] \Rightarrow [x_n \preceq x_m \ \wedge \ x_n \neq x_m].$$

Prove that there is no strictly decreasing sequence of elements in X.

EXERCISE 4.14. Prove that the last digit of $7^{7^{\cdot^{\cdot^{\cdot^{7}}}}}$ is 3 for any tower of 7s of height more than 1.

EXERCISE 4.15. Give another example that illustrates the need for a base case in a valid proof by induction.

EXERCISE 4.16. Assume that there is a polynomial of degree 4 in N that gives $\sum_{n=0}^{N} n^3$. Find the polynomial and then prove that the formula is correct by induction.

EXERCISE 4.17. Let $\mathbb{N}[x]$ be the set of polynomials with natural number coefficients. Define a relation $\preceq$ on $\mathbb{N}[x]$ as follows.

Let $p(x) = \sum_{n=0}^{N} a_n x^n$, and $q(x) = \sum_{n=0}^{M} b_n x^n$. Say that $p \preceq q$ if $a_k < b_k$, where k is the coefficient of highest degree at which p and q differ, or if $p = q$. Is $\preceq$ a linear ordering? Is it a well-ordering of $\mathbb{N}[x]$?

EXERCISE 4.18. Assume that there is a polynomial p of degree 5 such that

$$\sum_{n=0}^{N} n^4 = p(N).$$

Find p and prove that the formula you propose is correct.

EXERCISE 4.19. Determine the set of positive natural numbers n such that the sum of every n consecutive natural numbers is divisible by n.

EXERCISE 4.20. Let f be a real function such that, for $x, y \in \mathbb{R}$,

$$f(x + y) = f(x) + f(y).$$

Prove that

    (i) $f(0) = 0$

    (ii) $f(n) = nf(1)$

EXERCISE 4.21. Prove Corollary 4.9.

EXERCISE 4.22. Consider boxes with dimensions a, b, and c in which the sum of the dimensions (i.e., $a + b + c$) is fixed. Prove that the box with the largest possible volume has dimensions that satisfy $a = b = c$.

EXERCISE 4.23. Prove by induction that any well-formed propositional statement has a well-defined truth value.

EXERCISE 4.24. Prove by induction on the number of propositional connectives that every compound propositional statement is equivalent to a statement using only $\neg$ and $\vee$.

EXERCISE 4.25. Prove by induction on the number of propositional connectives that every compound propositional statement is equivalent to a statement using only $\neg$ and $\wedge$.

EXERCISE 4.26. Let Q_i be a quantifier for $1 \le i \le n$. For each Q_i, let Q_i^* be the complementary quantifier. That is, if $Q_i = \forall$, then $Q_i^* = \exists$; if $Q_i = \exists$, then let $Q_i^* = \forall$. Prove by induction on the number of quantifiers that

$$\neg(Q_1 x_1)(...)(Q_n x_n) P(x_1, \ldots, x_n) \equiv (Q_1^* x_1)(...)(Q_n^* x_n) \neg P(x_1, \ldots, x_n).$$

EXERCISE 4.27. Define the nth Fermat number to be

$$F_n := 2^{2^n} + 1, \qquad n \in \mathbb{N}.$$

(i) Show that the Fermat numbers satisfy

$$\prod_{k=0}^{n} F_k = F_{n+1} - 2.$$

(ii) Conclude that any two distinct Fermat numbers are coprime.

EXERCISE 4.28. Let $\langle a_n : n \in \mathbb{N} \rangle$ be a sequence of positive numbers. Suppose that $a_0 \le 1$, and that for all $N \in \mathbb{N}$,

$$a_{N+1} \le \sum_{n=0}^{N} a_n. \tag{4.20}$$

Prove

$$(\forall N \in \mathbb{N}) \, a_n \le 2^N. \tag{4.21}$$

EXERCISE 4.29. Let $\langle a_n : n \in \mathbb{N} \rangle$ be a sequence of positive numbers satisfying (4.20), and let $a_0 \le C$. What is the correct analogue of (4.21)? Prove your assertion.

EXERCISE 4.30. Let $\mathcal{F} = \{X_\alpha \mid \alpha \in A\}$ be an indexed family of pairwise disjoint sets. Suppose that each X_α is well-ordered by $\preceq_\alpha$ and that A is well-ordered by $\preceq$. Define a relation R on the union of all the sets in $\mathcal{F}$ by the following: for all $a, b \in \bigcup_{\alpha \in A} X_\alpha$, aRb iff
(a) $a \in X_{\alpha_1}$, $b \in X_{\alpha_2}$ and $\alpha_1 \prec \alpha_2$,

or

(b) $(\exists \alpha \in A)\ a, b \in X_\alpha$ and $a \preceq_\alpha b$.

Prove that R is a well-ordering of $\bigcup_{\alpha \in A} X_\alpha$.

EXERCISE 4.31. Let X be a finite set and $f : X \to X$. Prove that f is an injection iff f is a surjection.

CHAPTER 5

Limits

The idea of a limit is the cornerstone of calculus. It is somewhat subtle, which is why, although it was implicit in the work of Archimedes,[1] and essential to a proper understanding of Zeno's paradoxes, it took two thousand years to be understood fully. Calculus was developed in the seventeenth century by Newton and Leibniz with a somewhat cavalier approach to limits; it was not until the nineteenth century that a rigorous definition of limit was given, by Cauchy.

In Section 5.1 we define limits and prove some elementary properties. In Section 5.2 we discuss continuous functions, and in Section 5.3 we look at limits of sequences of functions.

5.1 Limits

Given a real function $f : X \to \mathbb{R}$, the intuitive idea of the statement

$$\lim_{x \to a} f(x) = L \qquad (5.1)$$

is that, as x gets closer and closer to a, the values of $f(x)$ get closer and closer to L. Making this notion precise is not easy—try to write down a mathematical definition now, before you read any further.

The idea behind the definition is to give a sequence of guarantees. Imagine yourself as an attorney trying to defend the claim (5.1). For

---

[1]Archimedes (287–212 BC) calculated the area under a parabola (what we would now call $\int_0^1 x^2 dx$) by calculating the area of the rectangles of width $1/N$ under the parabola and letting N tend to infinity. This is identical to the modern approach of finding an integral by taking a limit of Riemann sums.

concreteness, let us fix $g(x) = \frac{\sin(x)}{x}$ and try to defend the claim that

$$\lim_{x \to 0} g(x) = 1. \tag{5.2}$$

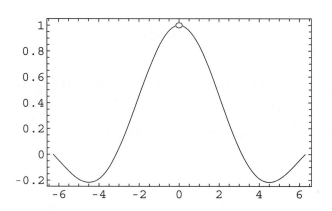

FIGURE 5.1 Plot of $\sin(x)/x$

The skeptical judge asks, "Can you guarantee that $g(x)$ is within 0.1 of 1?"

"Yes, your honor, provided that $|x| < 0.7$."

"Hmm, well can you guarantee that $g(x)$ is within 0.01 of 1?"

"Yes, your honor, provided that $|x| < 0.2$."

And so it goes. If, for every possible tolerance the judge poses, you can find a precision (i.e., an allowable deviation of x from a) that guarantees that the difference between the function value and the limit is within the allowable tolerance, then you can successfully defend the claim.

EXERCISE. Now try to give a mathematical definition of a limit, without reading any further.

We shall start with the case that the function is defined on an open interval.

DEFINITION.   Limit, $\lim\limits_{x \to a} f(x)$ Let I be an open interval and a some point in I. Let f be a real-valued function defined on $I \setminus \{a\}$. (It does not matter whether f is defined at a or not.) Then we say

$$\lim_{x \to a} f(x) \; = \; L$$

(in words, "the limit as x tends to a of $f(x)$ is L") if, for every $\varepsilon > 0$, there exists $\delta > 0$, so that

$$0 < |x - a| < \delta \quad \Longrightarrow \quad |f(x) - L| < \varepsilon. \tag{5.3}$$

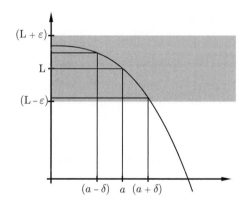

FIGURE 5.2 One choice of δ for a given ε

The condition $0 < |x - a| < \delta$ means we exclude $x = a$ from consideration. *Limits are about the behavior of a function near the point, not at the point.* For a function like $g(x) = \sin(x)/x$, the value at 0 is undefined; nevertheless $\lim_{x \to 0} g(x)$ exists and is the same as $\lim_{x \to 0}$ of the function

$$h(x) \; = \; \begin{cases} \sin(x)/x & x \neq 0 \\ 5 & x = 0. \end{cases}$$

REMARK. The use of ε for the allowable error and δ for the corresponding precision required is hallowed by long usage. Mathematicians need all the convenient symbols they can find. The Greek alphabet has long been used as a supplement to the Roman alphabet in Western mathematics, and you need to be familiar with it (see Appendix A for the Greek alphabet).

The main point to note in the definition is the order of the quantifiers: $\forall \varepsilon$, $\exists \delta$. What would it mean to say

$$(\exists \delta > 0) \; (\forall \varepsilon > 0) \quad [0 < |x - a| < \delta \quad \Longrightarrow \quad |f(x) - L| < \varepsilon] \; ?$$

To talk comfortably about limits, it helps to have some words that describe inequalities (5.3). Let us say that the ε-neighborhood of L is the set of points within ε of L, that is, the interval $(L - \varepsilon, L + \varepsilon)$. The punctured δ-neighborhood of a is the set of points within δ of a, excluding a itself, that is, $(a - \delta, a) \cup (a, a + \delta)$. When we speak of ε-neighborhoods and punctured δ-neighborhoods, we always assume that ε and δ are positive so that the neighborhoods are nonempty.

Then the definition of limit can be worded as "every ε–neighborhood of L has an inverse image under f that contains some punctured δ-neighborhood of a."

REMARK. We can revisit our courtroom analogy and say that to prove that f has limit L at a, we need a strategy that produces a workable δ for any ε. So a proof is essentially a function F that takes any positive ε and spits out a positive $\delta = F(\varepsilon)$ for which (5.3) works.

EXAMPLE 5.4. Let $f(x) = 5x + 2$. Prove $\lim_{x \to 3} f(x) = 17$.

Let $\varepsilon > 0$. We want to find a $\delta > 0$ so that the punctured δ-neighborhood of 3 is mapped into the ε-neighborhood of 17.

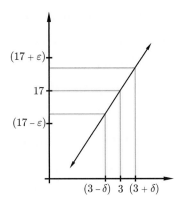

FIGURE 5.3 Relationship between δ and ε

Taking $\delta = \varepsilon/5$ will work, as will any smaller choice of δ. Indeed, if $0 < |x - 3| < \delta$, then $|f(x) - 17| < 5\delta = \varepsilon$.

EXAMPLE 5.5. This time, let $g(x) = 55x+2$. To prove $\lim_{x \to 3} g(x) = 167$, we must take $\delta \le \varepsilon/55$.

If two functions f and g both have limits at the point a, then so do all the algebraic combinations $f+g$, $f-g$, $f \cdot g$, and cf, for c a constant. The quotient f/g also has a limit at a, provided $\lim_{x \to a} g(x) \ne 0$. Moreover, these limits are what you would expect.

THEOREM 5.6. *Suppose f and g are functions on an open interval I, and at the point a in I both $\lim_{x \to a} f(x)$ and $\lim_{x \to a} g(x)$ exist. Let c be any real number. Then*

(i) $\quad \lim_{x \to a} [f(x) + g(x)] \quad = \quad \lim_{x \to a} f(x) + \lim_{x \to a} g(x)$

(ii) $\quad \lim_{x \to a} [f(x) - g(x)] \quad = \quad \lim_{x \to a} f(x) - \lim_{x \to a} g(x)$

(iii) $\quad \lim_{x \to a} cf(x) \quad = \quad c \left[\lim_{x \to a} f(x) \right]$

(iv) $\quad \lim_{x \to a} [f(x)g(x)] \quad = \quad \left[\lim_{x \to a} f(x) \right] \cdot \left[\lim_{x \to a} g(x) \right]$

(v) $\quad \lim_{x \to a} \frac{f(x)}{g(x)} \quad = \quad \dfrac{\lim_{x \to a} f(x)}{\lim_{x \to a} g(x)}, \qquad$ provided $\lim_{x \to a} g(x) \ne 0$

DISCUSSION. *How do we go about proving a theorem like this? Well, to start with, do not be intimidated by its length. Let us start on part (i). We only have the definition of limit to work with, so we only have one strategic option: prove directly that the definition is satisfied.*

PROOF OF (i). Let L_1 and L_2 be the limits of f and g respectively at a. Let ε be an arbitrary positive number. We must find a $\delta > 0$ so that

$$0 < |x - a| < \delta \quad \Longrightarrow \quad |f(x) + g(x) - (L_1 + L_2)| < \varepsilon.$$

The key idea, common to many limit arguments, is to use the observation that

$$|f(x) + g(x) - (L_1 + L_2)| \ \leq \ |f(x) - L_1| + |g(x) - L_2|. \qquad (5.7)$$

This is an application of the so-called *triangle inequality*, which you are asked later to prove (Lemma 5.11). It is the assertion that for any real numbers c and d, we have

$$|c + d| \ \leq \ |c| + |d|.$$

(What values of c and d yield (5.7)?) So if we can make *both* $|f(x) - L_1|$ and $|g(x) - L_2|$ small, then inequality (5.7) forces

$$|f(x) + g(x) - (L_1 + L_2)|$$

to be small too, which is what we want.

Since f and g have limits L_1 and L_2 at a, we know that there exist positive numbers δ_1 and δ_2 such that

$$0 < |x - a| < \delta_1 \quad \Longrightarrow \quad |f(x) - L_1| < \varepsilon$$
$$0 < |x - a| < \delta_2 \quad \Longrightarrow \quad |g(x) - L_2| < \varepsilon.$$

If $|x - a|$ is less than both δ_1 and δ_2, then both inequalities are satisfied, and we get

$$|f(x) + g(x) - (L_1 + L_2)| \ \leq \ |f(x) - L_1| + |g(x) - L_2| \ \leq \ \varepsilon + \varepsilon. \quad (5.8)$$

This is not quite good enough; we want the left-hand side of (5.8) to be bounded by ε, not 2ε. We are saved, however, by the requirement that for *any* positive number η, we can guarantee that f and g are in an η-neighborhood of L_1 and L_2, respectively. In particular, let η be $\varepsilon/2$. Since f and g have limits at a, there are positive numbers δ_3 and δ_4 so that

$$0 < |x - a| < \delta_3 \quad \Longrightarrow \quad |f(x) - L_1| < \frac{\varepsilon}{2}$$

$$0 < |x - a| < \delta_4 \quad \Longrightarrow \quad |g(x) - L_2| < \frac{\varepsilon}{2}.$$

So we set δ equal to the smaller of δ_3 and δ_4, and we get

$$0 < |x - a| < \delta \quad \Longrightarrow$$

$$|f(x) + g(x) - (L_1 + L_2)| \;\leq\; |f(x) - L_1| + |g(x) - L_2| \;<\; \varepsilon,$$

as required. $\triangleleft$

EXERCISE. Explain in words how the preceding proof worked. In shorthand, one could say that if F_1 and F_2 are strategies for proving $\lim_{x \to a} f(x) = L_1$ and $\lim_{x \to a} g(x) = L_2$ respectively, then

$$F \;=\; \min\left\{ F_1\left(\frac{\varepsilon}{2}\right),\, F_2\left(\frac{\varepsilon}{2}\right) \right\}$$

is a strategy for proving $\lim_{x \to a} f(x) + g(x) = L_1 + L_2$.

DISCUSSION. *What next? We could prove (ii) in a similar way, but mathematicians like shortcuts. Notice that if we prove (iii) and let $c = -1$, then we can apply (i) to $f + (-g)$ and get (ii) that way. Moreover, (iii) is just a special case of (iv), if we know that the constant function $g(x) = c$ has the limit c at every point. So let us prove (iv) next.*

PROOF OF (iv). Again, let ε be an arbitrary positive number. We must find a $\delta > 0$ so that

$$0 < |x - a| < \delta \quad \Longrightarrow \quad |f(x)g(x) - (L_1 L_2)| < \varepsilon.$$

It is not quite clear how close f and g have to be to L_1 and L_2 to conclude that their product is close enough to $L_1 L_2$, so let us play it safe by not choosing yet. For every $\varepsilon_1, \varepsilon_2 > 0$, we know there exist $\delta_1, \delta_2 > 0$ such that

$$0 < |x - a| < \delta_1 \qquad \Longrightarrow \qquad |f(x) - L_1| < \varepsilon_1$$
$$0 < |x - a| < \delta_2 \qquad \Longrightarrow \qquad |g(x) - L_2| < \varepsilon_2.$$

Now we use the second common trick in proving the existence of limits: add and subtract the same quantity so that one can factor.

$$
\begin{aligned}
|f(x)g(x) - L_1 L_2| &= |f(x)g(x) - L_1 g(x) + L_1 g(x) - L_1 L_2| \\
&\leq |f(x)g(x) - L_1 g(x)| + |L_1 g(x) - L_1 L_2| \\
&\leq |f(x) - L_1||g(x)| + |g(x) - L_2||L_1|. \qquad (5.9)
\end{aligned}
$$

Now if both summands on the last line can be made less than $\varepsilon/2$, we win. The second term is easy: we choose

$$\varepsilon_2 = \frac{\varepsilon}{2|L_1| + 1}.$$

Then there is a δ_2 so that

$$
\begin{aligned}
0 < |x - a| < \delta_2 \quad &\Longrightarrow \quad |g(x) - L_2| \quad < \varepsilon_2 \\
&\Longrightarrow \quad |g(x) - L_2||L_1| \quad < \frac{\varepsilon |L_1|}{2|L_1| + 1} < \frac{\varepsilon}{2}.
\end{aligned}
$$

(If $L_1 \neq 0$, we could have chosen $\varepsilon_2 = \frac{\varepsilon}{2|L_1|}$; we added 1 to the denominator just so we did not have to consider the two cases separately.)

What about the first summand in (5.9), the term $|f(x) - L_1||g(x)|$? First let us get some bound on how big $|g|$ can be. We know that if $0 < |x - a| < \delta_2$, then $|g(x) - L_2| < \varepsilon/(2|L_1| + 1)$, so

$$|g(x)| < |L_2| + \frac{\varepsilon}{2|L_1| + 1} =: M.$$

If we let $\varepsilon_1 = \varepsilon/(2M)$, we know that there exists $\delta_1 > 0$ so that

$$0 < |x - a| < \delta_1 \quad \Longrightarrow \quad |f(x) - L_1||g(x)| < \varepsilon_1 |g(x)|. \qquad (5.10)$$

Finally, we let $\delta = \min(\delta_1, \delta_2)$. For $0 < |x - a| < \delta$, both summands in (5.9) are less than $\varepsilon/2$: the second summand because $\delta \leq \delta_2$ and the first because when $0 < |x - a| < \delta$, inequality 5.10 is strengthened to

$$|f(x) - L_1||g(x)| < \varepsilon_1|g(x)| < \varepsilon_1 M = \varepsilon/2.$$

Therefore, for $0 < |x - a| < \delta$, we have $|f(x)g(x) - L_1L_2| < \varepsilon$, as desired. ◁

PROOF OF (iii). This is a special case of (iv), once we know that constant functions have limits. Let us state this as a lemma. Given Lemma 5.12, (iii) is proved and hence so is (ii).

PROOF OF (v). Exercise. □

LEMMA 5.11. *Triangle inequality* *Let c, d be real numbers. Then* $|c + d| \leq |c| + |d|$.

PROOF. Exercise. □

LEMMA 5.12. *Let $g(x) \equiv c$ be the constant function c. Then,*

$$(\forall a \in \mathbb{R}) \lim_{x \to a} g(x) = c.$$

PROOF. Exercise. □

EXAMPLE 5.13. The Heaviside function $H(t)$ is defined by

$$H(t) = \begin{cases} 0 & t < 0 \\ 1 & t \geq 0. \end{cases}$$

Show that H does not have a limit at 0.

DISCUSSION. *To prove that a limit does not exist, we must prove the opposite of $\forall \varepsilon \exists \delta$, that is, that $\exists \varepsilon \not\exists \delta$. As the gap between the function on $[0, \infty)$ and $(-\infty, 0)$ is 1, it is clear that any band of width < 1 cannot be wide enough to contain values of $H(t)$ for t on both sides of 0. So we will choose some $\varepsilon < 0.5$, and argue by contradiction.*

Suppose the limit exists and equals L. Let $\varepsilon = 1/4$. By hypothesis, there exists $\delta > 0$ such that

$$0 < |t| < \delta \implies |H(t) - L| < \frac{1}{4}.$$

But for t negative, this means $|L| < 1/4$; and for t positive, this means $|L - 1| < 1/4$. Thus we get a contradiction. □

If the function is defined on the closed interval $[c, d]$, we may still want to ask if it has a limiting value at c; if so, however, we only want to consider points near c that are in the domain of definition. More generally, we are led to the following definition of a restricted limit.

DEFINITION. Restricted limit, $\lim\limits_{X \ni x \to a} f(x)$ Suppose f is a real function and $X \subseteq \mathrm{Dom}(f)$. Let $a \in \mathbb{R}$. We say that $\lim\limits_{X \ni x \to a} f(x) = L$ if

$$(\forall \varepsilon > 0)\,(\exists \delta > 0)\,(\forall x \in X) \quad [0 < |x - a| < \delta] \implies |f(x) - L| < \varepsilon.$$

We read "$\lim\limits_{X \ni x \to a} f(x) = L$" as "the limit as x tends to a inside X of $f(x)$ is L." The following is an important special case of restricted limits.

DEFINITION. Right-hand limit, $\lim_{x \to a^+} f(x)$ Let $a, b, L \in \mathbb{R}$, $a < b$, and f be a real function defined on (a, b). We say that

$$\lim_{x \to a^+} f(x) = L$$

if

$$(\forall \varepsilon > 0)(\exists \delta > 0)\,[x \in (a, a + \delta)] \Rightarrow [|\,f(x) - L\,| < \varepsilon].$$

The number L is the right-hand limit of $f(x)$ at a. The left-hand limit is defined analogously. If $a, c, L \in \mathbb{R}$, $c < a$, and f is a real function defined on (c, a), we say that $\lim_{x \to a^-} f(x) = L$ if

$$(\forall \varepsilon > 0)(\exists \delta > 0)[x \in (a - \delta, a)] \Rightarrow [|\,f(x) - L\,| < \varepsilon].$$

Right-hand limits and left-hand limits are called *one-sided limits.* One-sided limits are examples of restricted limits.

EXAMPLE 5.14. Let $H(t)$ be the Heaviside function. Then

$$\lim_{t \to 0^+} H(t) = 1$$
$$\lim_{t \to 0^-} H(t) = 0.$$

5.2 Continuity

Most functions you have encountered have the property that at (almost) every point the function has a limit that agrees with its value there. This is a very useful feature of a function, and it is called *continuity.*

DEFINITION. Continuous Let f be a real function with domain $X \subseteq \mathbb{R}$. Let $a \in X$. Then we say f is continuous at a if $\lim_{X \ni x \to a} f(x) = f(a)$. We say f is continuous on X if it is continuous at every point of X.

Intuitively, the idea of a continuous function on an interval is that it has no jumps. We shall make this precise in Chapter 8 when we prove the Intermediate Value theorem 8.10, which asserts that if a continuous function on an interval takes on two distinct values c and d, it must also take on every value between c and d.

EXAMPLE 5.15. Prove that the function $f(x) = x^2$ is continuous on $\mathbb{R}$.

DISCUSSION. *How would we do this from first principles? We need to show that for every $a \in \mathbb{R}$, for every $\varepsilon > 0$, we can always find a $\delta > 0$ such that for any $x \in \mathbb{R}$*

$$|x - a| < \delta \quad \implies \quad |x^2 - a^2| < \varepsilon. \tag{5.16}$$

(Why don't we need to add the hypothesis $0 < |x - a|$?) The easiest way to do this is to write down a formula that, given a and ε, produces a δ satisfying (5.16).

PROOF. As $x^2 - a^2 = (x - a)(x + a)$, if $|x - a|$ is less than some number δ (still unspecified), then $|x^2 - a^2|$ is less than $\delta|x + a|$. So we want

$$\delta|x + a| \leq \varepsilon. \tag{5.17}$$

We cannot choose $\delta = \varepsilon/|x + a|$, because δ cannot depend on x. But if $|x - a| < \delta$, then

$$\begin{aligned} |x + a| &\leq |x| + |a| \\ &< |a| + \delta + |a| = 2|a| + \delta, \end{aligned}$$

so

$$|x^2 - a^2| < \delta(2|a| + \delta) \overset{?}{\leq} \varepsilon. \tag{5.18}$$

We must choose δ so that the last inequality holds. By the quadratic formula,

$$\delta(2|a| + \delta) \leq \varepsilon \iff \delta \leq \sqrt{|a|^2 + \varepsilon} - |a|.$$

So choose $\delta = \sqrt{|a|^2 + \varepsilon} - |a|$, and (5.16) holds. $\square$

REMARK. A formally correct proof could have been reduced to the following.

PROOF. Let $a \in \mathbb{R}$ and $\varepsilon > 0$. Then letting $\delta = \sqrt{|a|^2 + \varepsilon} - |a|$, we have $|x - a| < \delta \implies |x^2 - a^2| < \varepsilon$. QED.

However, while a diligent reader could verify that this proof is correct, pulling δ out of a hat like this does not give the reader the insight that our much longer proof does. Remember, a proof has more than one function: not only must it convince the reader that the claimed result is true, but it should also help the reader understand *why* the result is true. A good proof should be describable in a few English sentences so that a knowledgeable listener can then go write down a more detailed proof fairly easily.

REMARK. One does not need to choose the largest value of δ so that the inequality $\overset{?}{\leq}$ in (5.18) holds—any positive δ that satisfies the inequality will work. This allows one to simplify the algebra. For example, let δ_1 be such that

$$|x - a| < \delta_1 \;\Rightarrow\; |x + a| < 2|a| + 1.$$

(Such a δ_1 exists from the continuity of the simpler function $x \mapsto x$.) Then let

$$\delta = \min\left(\delta_1, \frac{\varepsilon}{2|a| + 1}\right)$$

and (5.17) holds.

One could imagine repeating proofs like the above to show that x^3, x^4, and so on are continuous, but we want to take big steps. Can we show all polynomials are continuous?

First observe that because limits behave well with respect to algebraic operations (Theorem 5.6) and continuity is defined in terms of limits, algebraic combinations of continuous functions are continuous.

PROPOSITION 5.19. *Suppose $f : X \to \mathbb{R}$ and $g : X \to \mathbb{R}$ are real functions that are continuous at $a \in X$. Let c and d be scalars.*[2] *Then $cf + dg$ and fg are both continuous at a and so is f/g if $g(a) \neq 0$.*

PROOF. Exercise.   □

Constant functions are continuous (Lemma 5.12), and the function $f(x) = x$ is continuous (Exercise 5.16). So one can prove by induction on the degree of polynomial, using Proposition 5.19, that all polynomials are continuous (Exercise 5.27). Once you have proved that all

---

[2]A scalar is just a fancy word for a number.

polynomials are continuous, you may prove that rational functions are continuous wherever the denominator does not vanish.

This result is used so frequently that we will state it formally.

PROPOSITION 5.20. *Every polynomial is continuous on* $\mathbb{R}$. *Every rational function is continuous wherever the denominator is nonzero.*

What about the exponential function

$$e^x := \sum_{n=0}^{\infty} \frac{x^n}{n!} \ ?$$

Each partial sum is a polynomial and hence continuous; so if we knew that the limit of a sequence of continuous functions were continuous, we would be done. This turns out, however, to be a subtle problem, which we address in the next section.

5.3 Sequences of Functions

An infinite sequence of numbers $\langle a_n \rangle$ tends to a limit L if a_n approaches L as n tends to infinity. Try to write down a formal definition of this before reading further.

DISCUSSION. *Hint: We have already seen how to encode the statement "approaches* L." *The difficulty is to encode "as* n *tends to infinity." How might you do this?*

DEFINITION. $\lim_{n \to \infty} a_n$, converge, diverge   The sequence $\langle a_n \rangle$ tends to the limit L as n tends to infinity, written

$$\lim_{n \to \infty} a_n \ = \ L,$$

if for every $\varepsilon > 0$ there exists $N \in \mathbb{N}$ such that

$$(\forall n \in \mathbb{N}) \ n > N \quad \implies \quad |a_n - L| \ < \ \varepsilon.$$

We say that the sequence $\langle a_n \rangle$ converges to L. If a sequence does not converge, we say it diverges.

EXAMPLE 5.21. Prove that the sequence

$$\langle \sin^2(n)/n \mid n \in \mathbb{N} \rangle$$

converges.

DISCUSSION. *It is generally easiest to prove that a sequence converges if we have an idea of its limit. To prove convergence of sequences without a candidate for the limit usually involves using the least upper bound property of $\mathbb{R}$ (which is covered in Chapter 8). It certainly seems that the terms in the sequence are getting closer to 0, so we try to show this rigorously.*

We observe that

$$(\forall n \in \mathbb{N}) \ \mid \sin^2(n) \mid \leq 1.$$

Hence

$$(\forall n \in \mathbb{N}) \ \mid \sin^2(n)/n \mid \leq \mid 1/n \mid .$$

Let $\varepsilon > 0$ and $N \in \mathbb{N}$ be such that $1/N \leq \varepsilon$. Then for any $n \geq N$,

$$\mid \sin^2(n)/n - 0 \mid \leq 1/n \leq \varepsilon.$$

Therefore

$$\lim_{n \to \infty} \frac{\sin^2(n)}{n} = 0.$$

EXAMPLE 5.22. For any $n \in \mathbb{N}$, let $a_n = (-1)^n$. Show that the sequence $\langle a_n \rangle$ diverges.

DISCUSSION. *Since the sequence alternates between -1 and 1, it is intuitively clear that the sequence does not tend to any particular number. We wish to show that a statement in the form*

$$(\exists L \in \mathbb{R})(\forall \varepsilon > 0)(\exists N \in \mathbb{N})(\forall n > N)(\ldots)$$

is false. So we must show that the negation of the statement is true. That is

$$(\forall L \in \mathbb{R})(\exists \varepsilon > 0)(\forall N \in \mathbb{N})(\exists n > N) \ \neg(\ldots).$$

For any $L \in \mathbb{R}$, if we pick $\varepsilon < 1$ we will not be able to capture both -1 and 1 in the ε-neighborhoods of L. This will prove that the sequence diverges.

Let $L \in \mathbb{R}$ and $\varepsilon < 1$. We show that for any $N \in \mathbb{N}$, there is $n > N$ such that
$$| (-1)^n - L | \geq \varepsilon.$$
Let $N \in \mathbb{N}^+$. We argue by cases.

Suppose $L < 0$. Then
$$| (-1)^{2N} - L | \geq 1 > \varepsilon.$$
Suppose $L \geq 0$. Then
$$| (-1)^{(2N+1)} - L | \geq 1 > \varepsilon.$$
Therefore the sequence $\langle a_n \rangle$ diverges.

EXAMPLE 5.23. For all $n \in \mathbb{N}$, let $a_n = \sum_{k=0}^{n} \left(\frac{k}{n} \right)^2 \frac{1}{n}$. What is the limit of the sequence $\langle a_n \rangle$?

DISCUSSION. *The terms of the sequence may be familiar to you as Riemann sums associated with the area under the parabola $f(x) = x^2$ between $x = 0$ and $x = 1$. We will use a combinatorial result we proved by induction in the last chapter.*

By Proposition 4.6,
$$\lim_{n \to \infty} \sum_{k=0}^{n} \left(\frac{k}{n} \right)^2 \frac{1}{n} = \lim_{n \to \infty} \left(\frac{1}{n^3} \right) \sum_{k=0}^{n} k^2$$
$$= \lim_{n \to \infty} \left(\frac{1}{n^3} \right) \frac{(n)(n+1)(2n+1)}{6}$$
$$= 1/3.$$

Verification of the last equality is left to the reader as an exercise.

In the next section we are particularly interested in infinite sums.

DEFINITION. Infinite sum, Partial sum, $\sum_{k=0}^{\infty} a_k$ Let $\langle a_k \mid k \in \mathbb{N} \rangle$ be a sequence of numbers. The nth partial sum of the sequence is

$$s_n = \sum_{k=0}^{n} a_k.$$

The infinite sum of the sequence is

$$\sum_{k=0}^{\infty} a_k := \lim_{n \to \infty} s_n.$$

The infinite sum is the limit of the sequence of *partial* sums, $\langle s_n \rangle$.

EXAMPLE 5.24. Show that

$$\sum_{k=0}^{\infty} \frac{1}{2^k} = 2.$$

Let s_n be the nth partial sum. We need to show that

$$\lim_{n \to \infty} s_n = 2.$$

Let $\varepsilon > 0$ and $N \in \mathbb{N}$ be such that $\frac{1}{2^N} < \varepsilon$. We show that if $n \geq N$, then

$$|\, s_n - 2 \,| < \varepsilon.$$

Since the series $\sum_{k=0}^{\infty} \frac{1}{2^k}$ is geometric, we know that

$$s_n = \frac{1 - 2^{-(n+1)}}{1 - 1/2} = 2 - 2^{-n}.$$

So if $n \geq N$, then

$$|\, s_n - 2 \,| = 2^{-n} < \varepsilon.$$

In analysis, one is often concerned with a sequence of functions f_n. For example, f_n might be the nth-order Taylor polynomial of some function f, and one wants to know whether this sequence f_n converges to f; or the sequence f_n may represent functions whose graphs have a fixed boundary curve in $\mathbb{R}^3$ and have decreasing areas, and one wants to know if the sequence converges to the graph of a function with minimal area for that boundary. This sort of problem is so important that mathematicians study different ways in which a sequence of functions might converge. The most obvious way is pointwise.

DEFINITION.   Pointwise convergence   A sequence of functions f_n on a set X converges pointwise to the function f if, for all x in X, the sequence of numbers $\langle f_n(x) \rangle$ converges to $f(x)$.

In order for the definition to make sense, we require that

$$X \subseteq \bigcap_{n \in \mathbb{N}} \mathrm{Dom}(f_n).$$

If $a \in X$, the pointwise convergence of a sequence of functions, $\langle f_n \rangle$, at the point a is dependent on the convergence of the sequence of numbers, $\langle f_n(a) \rangle$. If you do not understand convergence of a sequence of numbers, you cannot understand convergence of a sequence of functions.

EXAMPLE 5.25. Consider the functions $f_n(x) = x^n$. On the open interval $(-1, 1)$, these functions converge pointwise to 0. At the point 1, the functions converge to 1; at the point -1, the functions do not converge, because the values oscillate between $+1$ and -1. Outside of the set $(-1, 1]$ the sequence of functions diverges.

The preceding example illustrates the main problem with pointwise convergence: the sequence of continuous functions x^n on the set $[0, 1]$ converges, but the function to which it converges is not continuous. Even Cauchy made this mistake: he stated as a theorem in his book *Cours d'Analyse* (1821) that if a sequence of continuous functions converges pointwise, then its limit is continuous.[3] To get around this problem, we introduce the notion of *uniform convergence*.

---

[3]See the book [4] by Imre Lakatos for an interesting historical discussion of Cauchy's mistake and the discovery of uniform convergence, independently, by Seidel and Stokes in 1847.

DEFINITION. Uniform convergence The sequence of real functions f_n defined on a set $X \subseteq \mathbb{R}$ is said to converge uniformly to the function f on X if, for every $\varepsilon > 0$, there exists $N \in \mathbb{N}$ such that, for every x in X, whenever $n > N$ then $|f_n(x) - f(x)| < \varepsilon$. In logical notation, this reads:

$$(\forall \varepsilon > 0) \ (\exists N \in \mathbb{N}) \ (\forall x \in X) \ (\forall n > N) \quad |f_n(x) - f(x)| < \varepsilon.$$

Note the big difference between pointwise and uniform convergence: in pointwise convergence N can depend on x; in uniform convergence it cannot. The importance of uniform convergence stems from the following theorem.

THEOREM 5.26. *Let f_n be a sequence of continuous functions on X that converges uniformly to f on X. Then f is continuous on X.*

DISCUSSION. *We must show $|f(x) - f(a)|$ is small when x is close to a. We know that $|f_n(x) - f(x)|$ is small for all x; so we refine the trick from page 142 and add and subtract the same thing twice, writing*

$$f(x) - f(a) \ = \ [f(x) - f_n(x)] \ + \ [f_n(x) - f_n(a)] \ + \ [f_n(a) - f(a)].$$

Then we try to make each of the three grouped pairs small, so their sum is small. This is sometimes called an $\varepsilon/3$ argument because if we make each term smaller than $\varepsilon/3$, then their sum is smaller than ε.

PROOF. Fix some point $a \in X$, and let $\varepsilon > 0$. We must find $\delta > 0$ so that

$$|x - a| < \delta \quad \Longrightarrow \quad |f(x) - f(a)| < \varepsilon. \tag{5.27}$$

To do this, we split $f(x) - f(a)$ into three parts:

$$f(x) - f(a) \ = \ [f(x) - f_n(x)] \ + \ [f_n(x) - f_n(a)] \ + \ [f_n(a) - f(a)].$$

Choose N so that $n \geq N$ implies $|f_n(x) - f(x)| < \varepsilon/3$ for all x. Choose $\delta > 0$ so that $|f_N(x) - f_N(a)| < \varepsilon/3$ whenever $|x - a| < \delta$. Then by the triangle inequality, for $|x - a| < \delta$, we have

$$
\begin{aligned}
|f(x) - f(a)| &\leq |f(x) - f_N(x)| + |f_N(x) - f_N(a)| + |f_N(a) - f(a)| \\
&\leq \frac{\varepsilon}{3} + \frac{\varepsilon}{3} + \frac{\varepsilon}{3} = \varepsilon.
\end{aligned}
$$

$\square$

QUESTION. Where did we use the hypothesis that the convergence was uniform?

We can use Theorem 5.26, for example, to prove that the exponential function is continuous. We consider the exponential function as its Taylor series

$$
e^x = \sum_{k=0}^{\infty} \frac{x^k}{k!}.
$$

Recall that the expression $\sum_{k=0}^{\infty} \frac{x^k}{k!}$ is a shorthand for

$$
\lim_{n \to \infty} \sum_{k=0}^{n} \frac{x^k}{k!}.
$$

For any real number a, $\sum_{k=0}^{\infty} \frac{a^k}{k!}$ is an infinite sum which converges if its corresponding sequence of partial sums converges. By the ratio test, the exponential series converges for all real a. (For a formal proof of the ratio test, see Theorem 8.9.)

PROPOSITION 5.28. *The exponential function is continuous on* $\mathbb{R}$.

PROOF. Let

$$
p_n(x) := \sum_{k=0}^{n} \frac{x^k}{k!}
$$

be the nth-order Taylor polynomial. We know each p_n is continuous, by Proposition 5.20. If we knew that $p_n(x)$ converged uniformly to e^x, we would be done with the help of Theorem 5.26.

It is not true that p_n converges uniformly on $\mathbb{R}$ (why?). However, the sequence does converge uniformly on every interval $[-R, R]$, and this is good enough to conclude that e^x is continuous on $\mathbb{R}$ (why?).

To see this latter assertion, fix $R > 0$ and $\varepsilon > 0$. We must find N so that for all $n > N$ and all $x \in [-R, R]$, we have $|e^x - p_n(x)| < \varepsilon$. Notice that

$$|e^x - p_n(x)| \; = \; \left| \frac{x^{n+1}}{(n+1)!} + \frac{x^{n+2}}{(n+2)!} + \cdots \right|.$$

For each n, the right-hand side is maximized on $[-R, R]$ by its value at R (why?); and as n increases, this remainder decreases monotonically (because you lose more and more positive terms). As we know the exponential series for e^R converges, choose an N so that $e^R - p_N(R)$ is less than ε. Then for all x in $[-R, R]$ and all $n \geq N$, we have $|e^x - p_n(x)| < \varepsilon$, as desired. $\qquad\qquad\square$

The sine and cosine functions can be defined in terms of their Taylor series too:

$$\sin(x) \quad := \quad \sum_{n=0}^{\infty} (-1)^n \frac{x^{2n+1}}{(2n+1)!}$$

$$\cos(x) \quad := \quad \sum_{n=0}^{\infty} (-1)^n \frac{x^{2n}}{(2n)!}.$$

They can be proved to be continuous by similar arguments.

REMARK. Notice that in our definitions of limits and continuity, we are using the absolute value just to measure distances. In other words, we are saying that f is continuous at a if, for all $\varepsilon > 0$, we can find $\delta > 0$, such that whenever the distance from x to a is less than δ, then the distance from $f(x)$ to $f(a)$ is less than ε. This definition makes perfectly good sense whenever one has a way of measuring distances on the domain and codomain. For example, if the function maps $\mathbb{R}^m$ to $\mathbb{R}^n$, one can measure distances in the usual Euclidean way. In even greater generality, mathematicians use something called *metrics* to measure distances, and once one has metrics, one can discuss the continuity of functions in a similar way to our discussion for real functions.

The mathematics of this chapter—a close look at the behavior of real functions—is called *analysis*. This comprises one of the three major disciplines of pure mathematics; the other two are geometry and algebra. A good introduction to analysis is Walter Rudin's book [**7**].

5.4 Exercises

EXERCISE 5.1. Prove that the definitions of limit on pages 137 and 138 are the same.

EXERCISE 5.2. Prove Lemma 5.11 and the related assertion that $|c| - |d| \leq |c + d|$.

EXERCISE 5.3. For $n \in \mathbb{N}^+$, $a_i \in \mathbb{R}$ $(1 \leq i \leq n)$, prove that

$$\left| \sum_{i=1}^{n} a_i \right| \leq \sum_{i=1}^{n} |a_i| .$$

EXERCISE 5.4. Prove Lemma 5.12.

EXERCISE 5.5. Prove part (v) of Theorem 5.6.

EXERCISE 5.6. Give an example of two functions f and g that do not have limits at a point a but such that $f + g$ does. For the same pair of functions, can $f - g$ also have a limit at a?

EXERCISE 5.7. Assume that f is a real function and $\lim_{x \to a} f(x) = L$. Prove that if $X \subseteq \text{Dom}(f)$, then

$$\lim_{X \ni x \to a} f(x) = L.$$

EXERCISE 5.8. Use Archimedes's method (the method of Riemann sums) to prove that

$$\int_0^1 x^2 dx = \frac{1}{3}.$$

(You will need to know a formula for $\sum_{k=0}^{n} k^2$—see Proposition 4.6).

EXERCISE 5.9. Use Archimedes's method to prove that

$$\int_0^1 x^3 dx = \frac{1}{4}.$$

(See Exercise 4.16).

EXERCISE 5.10. Prove that the Heaviside function has both left- and right-hand limits at 0.

EXERCISE 5.11. Prove that a function has a limit at a point if and only if it has both left and right limits at that point and their values coincide.

EXERCISE 5.12. Prove that Theorem 5.6 applies to restricted limits.

EXERCISE 5.13. The point a is a *limit point* of the set X if, for every $\delta > 0$, there exists a point x in $X \setminus \{a\}$ with $|x - a| < \delta$. Let f be a real-valued function on $X \subseteq \mathbb{R}$. Prove that if a is a limit point of X, then, if f has a restricted limit at a, it is unique. Prove that if a is not a limit point of X, then every real number is a restricted limit of f at a.

EXERCISE 5.14. Prove that $\lim_{x \to 0} \sin(x)/x = 1$.

EXERCISE 5.15. Prove Proposition 5.19.

EXERCISE 5.16. Prove that the function $f(x) = x$ is continuous everywhere on $\mathbb{R}$.

EXERCISE 5.17. A formula for the Fibonacci numbers is given in Exercise 4.12. Evaluate $\lim_{n \to \infty} F_{n+1}/F_n$.

EXERCISE 5.18. How large must n be to ensure that F_{n+1}/F_n is within 10^{-1} of the limit in Exercise 5.17? Within 10^{-2}? Within 10^{-k}?

EXERCISE 5.19. Define the function $\psi : \mathbb{R} \to \mathbb{R}$ by

$$\psi(x) := \begin{cases} 0 & x \notin \mathbb{Q} \\ 1 & x \in \mathbb{Q}. \end{cases}$$

Prove that ψ is discontinuous everywhere.

EXERCISE 5.20. Define the function $\phi : \mathbb{R} \to \mathbb{R}$ by

$$\phi(x) := \begin{cases} 0 & x \notin \mathbb{Q} \\ \frac{1}{n} & x \in \mathbb{Q} \setminus \{0\}, \ x = \frac{m}{n}, \ \gcd(m,n) = 1, \ n > 0 \\ 1 & x = 0. \end{cases}$$

Prove that ϕ is continuous at every irrational number and discontinuous at every rational number.

EXERCISE 5.21. Prove that a real-valued function f on an open interval I is continuous at any point where its derivative exists, that is, where

$$\lim_{x \to a} \frac{f(x) - f(a)}{x - a}$$

exists. What is the converse of this assertion? Prove that the converse is not true.

EXERCISE 5.22. Prove that if the function f has the limit L from the right at a, then the sequence $f(a + \frac{1}{n})$ has limit L as $n \to \infty$. Show that the converse is false in general.

EXERCISE 5.23. Let f and g be real functions. Let $a \in \mathbb{R}$ and suppose that

$$\lim_{x \to a} g(x) = L_1$$

and

$$\lim_{x \to L_1} f(x) = L_2.$$

Prove that

$$\lim_{x \to a} f \circ g = L_2.$$

If g is continuous at a and f is continuous at $g(a)$, is $f \circ g$ continuous at a?

EXERCISE 5.24. Let f be a real function, $a \in \mathbb{R}$ and $\lim_{x \to a} f(x) = L$. If $\langle a_n \rangle$ converges to a, prove that $\langle f(a_n) \rangle$ converges to L.

EXERCISE 5.25. Complete Example 5.23. That is, prove that

$$\lim_{n \to \infty} \left(\frac{1}{n^3} \right) \frac{(n)(n+1)(2n+1)}{6} = \frac{1}{3}.$$

EXERCISE 5.26. Evaluate

$$\lim_{n \to \infty} \sum_{k=0}^{n} \left(\frac{k}{n} \right) \frac{1}{n}.$$

Can you give a geometrical interpretation of this limit?

EXERCISE 5.27. Use induction to prove that every polynomial is continuous at every real number.

EXERCISE 5.28. Let $-1 < x < 1$. Prove that the geometric series with ratio x, $\sum_{k=0}^{\infty} x^k$, converges to $\frac{1}{1-x}$.

EXERCISE 5.29. Let the Fibonacci numbers F_n be defined as in Exercise 4.12. Consider the power series $F(x) = \sum_{n=1}^{\infty} F_n x^n$. Prove that the power series satisfies

$$F(x) = x^2 F(x) + x F(x) + x. \tag{5.29}$$

Solve (5.29) for $F(x)$, decompose it by partial fractions, and use Exercise 5.28 to derive Formula 4.19. This technique to find a formula for F_n by studying the function F is often fruitful. The function F is called the *generating function* for the sequence.

EXERCISE 5.30. Suppose one defines a sequence with the same recurrence relation as the Fibonacci numbers, $F_{n+2} = F_{n+1} + F_n$, but with different starting values for F_1 and F_2. Find the generating function for the new sequence, and hence calculate a formula for the general term. Is $\lim_{n \to \infty} F_{n+1}/F_n$ always the same?

EXERCISE 5.31. Prove that sine and cosine are continuous functions on all of $\mathbb{R}$.

CHAPTER 6

Cardinality

In this chapter we use functions to explore the idea of the size of a set. The results we derive are deep and very interesting, especially when we consider the simplicity of the tools we are using. Of course, we shall have to use these tools somewhat cleverly.

Set theory comes in different flavors. The most difficult is axiomatic set theory. Many interesting results have been derived in formal axiomatic set theory, but the topic is advanced and not suitable for an introduction to higher mathematics. Instead, we shall study what is called naive set theory. The use of the word "naive" is not pejorative but is meant to differentiate this approach from axiomatic set theory. Most mathematicians have studied naive set theory, but relatively few have worked extensively with set axioms.

6.1 Cardinality

We wish to compare the size of sets. The fundamental tool for our investigation is the bijection. In the case of finite sets, which can be exhaustively listed, this is easy. Given any two finite sets, X and Y, we could list the elements and count them. Provided that our lists have no redundancies, the larger set is the one with the higher count. The act of listing the elements in a set, where this is possible, is also defining a bijection from a natural number (interpreted as a set) to the set being counted. The idea of using functions to compare the size of sets can be generalized to arbitrary sets.

When it comes to comparing the size of infinite sets there are competing intuitions. On the one hand we have an intuition that if one set is a proper subset of another set, it should be smaller. On the other hand if two sets are infinite, how can one be larger than the other? Using bijections, injections, and surjections to define the relative size of sets allows us to see our way through this paradox.

DEFINITION.   Equinumerous, cardinality   Let X and Y be sets. We say that X and Y have the same cardinality if there is a bijection $f : X \rightarrowtail Y$. We can express that two sets have the same cardinality by

$$| \, X \, | = | \, Y \, | \, .$$

If $| \, X \, | = | \, Y \, |$, then we say that X and Y are equinumerous.

CLAIM. Equinumerosity is an equivalence relation.

(Prove this; see Exercise 6.2.)

Although we used the ideas of finite and infinite before now, we shall define the ideas in terms of bijections.

DEFINITION.   Finite, infinite   Let X be a set. X is finite if there exists some $n \in \mathbb{N}$ and a bijection $f : \ulcorner n \urcorner \rightarrowtail X$. In the case that $X = \emptyset$, we say that X is bijective with $\ulcorner 0 \urcorner$ via the empty function. If X is not finite, we say that X is infinite.

So a set is finite if it is bijective with a set $\ulcorner n \urcorner$ for some $n \in \mathbb{N}$. It is probably no surprise that a set cannot be bijective with different natural numbers.

PROPOSITION 6.1. *Let* $m, n \in \mathbb{N}$. *Then*

$$(\ | \ulcorner m \urcorner | = | \ulcorner n \urcorner | \ ) \ \Longleftrightarrow \ (\ m = n \ ).$$

DISCUSSION. *We prove the nontrivial direction of this biconditional by induction on one of the integers in the statement.*

PROOF. $\Leftarrow$

Let $m = n$. Then it is obvious that

$$| \ulcorner m \urcorner | = | \ulcorner n \urcorner | .$$

$\Rightarrow$

We argue by induction on m.

BASE CASE:

If $m = 0$ and $| \ulcorner n \urcorner | = | \ulcorner m \urcorner |$, then clearly $n = 0$.

INDUCTION STEP:

Let $m \in \mathbb{N}$ and assume that

$$(\forall n \in \mathbb{N}) \; [\; | \ulcorner m \urcorner | = | \ulcorner n \urcorner | \;] \Rightarrow [m = n].$$

We show that

$$(\forall n \in \mathbb{N}) \; [\; | \ulcorner m + 1 \urcorner | = | \ulcorner n \urcorner | \;] \Rightarrow [m + 1 = n].$$

Assume that

$$| \ulcorner m + 1 \urcorner | = | \ulcorner n \urcorner | .$$

Let

$$f : \ulcorner m + 1 \urcorner \rightarrowtail \ulcorner n \urcorner.$$

DISCUSSION. *A natural way to proceed with this argument is to restrict the domain of f to $\ulcorner m \urcorner$ and use the induction hypothesis. Unfortunately if $f(m) \neq n - 1$ then $f|_{\ulcorner m \urcorner}$ is not a bijection from $\ulcorner m \urcorner$ to $\ulcorner n - 1 \urcorner$, and the induction hypothesis will not directly apply. To address this issue, we shall define a permutation $g : \ulcorner m+1 \urcorner \rightarrow \ulcorner m+1 \urcorner$ that rearranges the elements of $\ulcorner m + 1 \urcorner$ so that $f \circ g$ will be a bijection satisfying*

$$(f \circ g)(m) = n - 1.$$

We define $g : \ulcorner m + 1 \urcorner \rightarrow \ulcorner m + 1 \urcorner$ as follows:

$$g(x) = \begin{cases} f^{-1}(n - 1) & \text{if } x = m \\ m & \text{if } x = f^{-1}(n - 1) \\ x & \text{otherwise.} \end{cases}$$

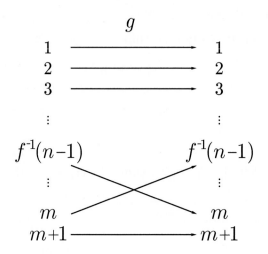

FIGURE 6.1 Picture of the permutation g

Let $h = f \circ g$. Then h is a bijection and

$$h(m) = (f \circ g)(m) = n - 1.$$

Therefore

$$h|_{\ulcorner m \urcorner} : \ulcorner m \urcorner \rightarrowtail \ulcorner n - 1 \urcorner.$$

By the induction hypothesis,

$$m = n - 1.$$

Therefore

$$m + 1 = n.$$

By the induction principle,

$$(\forall m \in \mathbb{N})(\forall n \in \mathbb{N}) \left(|\ulcorner m \urcorner| = |\ulcorner n \urcorner| \right) \Rightarrow (m = n).$$

$\square$

COROLLARY 6.2. *If X is a finite set, there is exactly one $n \in \mathbb{N}$ such that $\ulcorner n \urcorner$ is bijective with X.*

DISCUSSION. *This is a standard uniqueness argument. We assume that a set is bijective with natural numbers $\ulcorner n \urcorner$ and $\ulcorner m \urcorner$, and we use that the composition of bijections is a bijection to show that $m = n$. This is not a proof by contradiction. Rather we are proving that any two names for natural numbers that are bijective with X must name the same natural number.*

PROOF. X is finite, so there is $n \in \mathbb{N}$ such that

$$| X | = | \ulcorner n \urcorner | .$$

Let $m \in \mathbb{N}$ and

$$| X | = | \ulcorner m \urcorner | .$$

Let $f : X \rightarrowtail \ulcorner n \urcorner$ and $g : X \rightarrowtail \ulcorner m \urcorner$. Then $g^{-1} : \ulcorner m \urcorner \rightarrowtail X$. The composition of bijections is a bijection, so

$$f \circ g^{-1} : \ulcorner m \urcorner \rightarrowtail \ulcorner n \urcorner.$$

By Proposition 6.1,

$$m = n.$$

$\square$

DEFINITION. Finite cardinality If X is a finite set, we say that it has finite cardinality. Let $n \in \mathbb{N}$ be the unique natural number such that $\ulcorner n \urcorner$ is bijective with X. Then we say that X has cardinality n, or

$$| X | = n.$$

Corollary 6.2 guarantees that the cardinality of a finite set is well-defined.

6.2 Infinite Sets

Infinite sets are mysterious. Many classical paradoxes address historical confusions about the idea of infinity. At the same time, mathematicians from the ancient Greeks on have found it impossible to develop mathematical thinking without the use of infinity. Why is this so? From a metaphysical point of view, the idea of infinity is probably not necessary. From a physical point of view, there is no evidence for infinity. That is, the universe, as we understand it, is finite. Even from a theological point of view, infinity is to some extent the complement of the finite—and correspondingly gives rise to its own paradoxes.

Infinity has troubled some mathematicians and philosophers, and a few have tried to dispense with it. There are not many adherents to this school. The idea of infinity is so useful that the mathematics student will have to develop some comfort with the idea—and its logical consequences. At any rate, infinity clearly exists in the mathematical universe, whether or not it exists in the natural world, and using infinity has been crucial to developing a mathematical understanding of the natural world. In this section we begin an investigation of infinite sets.

We shall use injections and surjections to build some analytical machinery for comparing sets.

NOTATION. $\preceq$ Let X and Y be sets. We write $X \preceq Y$ if there is an injection

$$f : X \to Y.$$

This notation suggests that, under the conditions of the definition, we think of X as being "no bigger than" than Y. This makes sense, since we are able to associate to any element of X a distinct element of Y. If f in the definition is a surjection, then f is also a bijection and $|X| = |Y|$. Otherwise, f is a function that associates with each element of the range of f (which is a proper subset of Y) a unique element of X, and Y still has some elements unaccounted for by f. So Y might be "bigger" than X, but it certainly will not be "smaller." You might wish to consider this definition in the special case of finite sets X and Y. You will observe that

$$X \preceq Y \iff |X| \leq |Y|.$$

In Exercise 6.3 you are asked to prove that $\preceq$ is transitive and reflexive.

REMARK. Are any two sets comparable with respect to $\preceq$? Rather surprisingly, it requires a more advanced assumption, called the Axiom of Choice (see Appendix B), in order to guarantee the comparability of all pairs of sets. Virtually all mathematicians accept the Axiom of Choice. We shall assume the Axiom of Choice in this text.

If $X \preceq Y$ and $Y \preceq X$, we would hope that X and Y are the same size. This is indeed true, though the proof is a little tricky. The result is very useful because it is often much easier to write down two injections than one bijection.

THEOREM 6.3. *Schröder-Bernstein theorem* *Let X and Y be sets. If $X \preceq Y$ and $Y \preceq X$, then $|X| = |Y|$.*

DISCUSSION. *The idea behind this proof is as follows. We show that $|X| = |Y|$ by constructing a bijection $F : X \rightarrowtail Y$. We are given injections $f : X \rightarrow Y$ and $g : Y \rightarrow X$. We build F using the injections f and g as guides. That is, we wish to define F so that for each $x \in X$, either $F(x) = f(x)$ or $F(x) = g^{-1}(x)$. It is obvious that this cannot be accomplished blindly. For instance, if $x \in X \setminus g[Y]$, our hand is forced, and $F(x) = f(x)$. Similarly, if $y \in Y \setminus f[X]$, our only chance of achieving our objective is for $F(g(y)) = y$. If we make the wrong choice for $F(x)$, we shall lose the use of f and g as guides. We might consider F undecided about x since f and g^{-1} do not agree. The solution is to use f and g to move back and forth between X and Y until we find that our hand is forced.*

PROOF. Let

$$f : X \rightarrow Y$$

and

$$g : Y \to X$$

be injections. We may assume that X and Y are disjoint.

DISCUSSION. *If X and Y are not disjoint, we can replace X with $X \times \{0\}$ and Y with $Y \times \{1\}$. The existence of a bijection*

$$g : X \times \{0\} \rightarrowtail Y \times \{1\}$$

clearly implies the existence of a bijection from X to Y.

If $x \in X$ we say $y \in Y$ is the predecessor of x if $g(y) = x$. Analogously, if $y \in Y$ we say that $x \in X$ is the predecessor of y if $f(x) = y$. It is possible for an element not to have a predecessor. For example, if $x \in X \setminus g[Y]$, then x has no predecessor. However, if an element does have a predecessor, that predecessor is unique (since f and g are both injections).

Let $w, z \in X \cup Y$. We say that z is an antecedent of w if there is a finite sequence $\langle z_n \mid 0 \leq n \leq N \rangle$ for some $N \geq 1$ satisfying

(1) $z = z_0$
(2) $w = z_N$
(3) For $n < N$, z_n is the predecessor of z_{n+1}

The sequence alternates between elements of X and elements of Y. Let $w \in X \cup Y$ and let $A(w)$ be the set of antecedents of w. For all such w, there are three mutually exclusive possibilities:

(1) The set $A(w)$ is finite and has an even number of elements (or is empty).
(2) The set $A(w)$ is finite and has an odd number of elements.
(3) The set $A(w)$ is infinite.

Let

$$X_e = \{x \in X \mid A(x) \text{ is even or } A(x) = \emptyset\}$$

and

$$X_o = \{x \in X \mid A(x) \text{ is odd }\}.$$

Also, let

$$Y_o = \{y \in Y \mid A(y) \text{ is odd }\}$$

and

$$Y_e = \{y \in Y \mid A(y) \text{ is even or } A(y) = \emptyset\}.$$

Finally let $X_i \subseteq X$ be the set of elements of X with infinitely many antecedents, and let $Y_i \subseteq Y$ be the set of elements of Y with infinitely many antecedents. The collection

$$\{X_e, X_o, X_i\}$$

is obviously a partition of X. Similarly,

$$\{Y_o, Y_e, Y_i\}$$

is a partition of Y.

We are now in a position to define a bijection between X and Y. Let

$$F(x) = \begin{cases} f(x) & \text{if} & x \in X_i \\ f(x) & \text{if} & x \in X_e \\ g^{-1}(x) & \text{if} & x \in X_o. \end{cases}$$

DISCUSSION. *We have some work left in this proof. We need to verify that F is a bijection from X to Y. The idea behind the definition of F may not be obvious, so let us investigate the motivation for the definition. Suppose that f and g fail to be surjections (if either of the functions is a surjection there would be nothing to prove, since it would also be a bijection). Let $x \in X \setminus g[Y]$ and $y \in Y \setminus f[X]$. Since $x \notin g[Y]$, the only possible choice for $F(x)$ is $f(x)$. Similarly, $y \notin f[X]$, and the only possible value of $F^{-1}(y)$ is $g(y)$. But this does not solve all of our problems. The set $X \setminus g[Y]$ is made up of those members of X that have no predecessors, and $Y \setminus f[X]$ is composed of the members of Y with no predecessors. If we want to define F by piecing together f*

and g, we have found that our hands were forced with these sets. Now suppose that $x \in X$ has exactly one antecedent. Then $g^{-1}(x)$ has no predecessor. As we observed earlier, we need to satisfy

$$F^{-1}(g^{-1}(x)) \;=\; g(g^{-1}(x)) \;=\; x$$

and therefore we must satisfy

$$F(x) \;=\; g^{-1}(x).$$

Similarly, if $y \in Y$ has exactly one antecedent, we must satisfy

$$F^{-1}(y) \;=\; f^{-1}(y).$$

If an element w of $X \cup Y$ has finitely many antecedents, $F|_{A(w)}$ will be determined by the constraint imposed by the antecedent with no predecessor.

We claim that

$$F : X \rightarrowtail Y.$$

It is easily seen that F is well-defined. Since $X_o \subseteq g[Y]$ and g is an injection, $F|_{X_o} = g^{-1}|_{X_o}$ is well-defined. That F is well-defined on X_e and X_i is obvious. Furthermore

$$\begin{aligned} F[X_e] &= f[X_e] = Y_o \\ F[X_o] &= g^{-1}[X_o] = Y_e \end{aligned}$$

and

$$F[X_i] \;=\; f[X_i] \;=\; Y_i.$$

DISCUSSION. *Although we had no choice in the definition of F on X_e and X_o, we could have defined F so that $F|_{X_i} = g^{-1}|_{X_i}$.*

Therefore,

$$\begin{aligned} F[X] &= F[X_e \cup X_o \cup X_i] \\ &= f[X_e] \cup g^{-1}[X_o] \cup f[X_i] = Y_o \cup Y_e \cup Y_i = Y. \end{aligned}$$

So F is a surjection. We show that F is an injection. Let $x, z \in X$, and suppose $F(x) = F(z)$. If $x \in X_e$, then $F(x) \in Y_o$ and $z \in X_e$. Hence

$$F(x) = f(x) = f(z) = F(z).$$

Since f is an injection, so is $f|_{X_e}$. Therefore $x = z$.

If $x \in X_o$, then $F(x) \in Y_e$ and $z \in X_o$. So

$$F(x) = g^{-1}(x) = g^{-1}(z) = F(z).$$

The function g is an injection, therefore $g^{-1}|_{X_o}$ is an injection and so $x = z$.

Finally, if $x \in X_i$, then $F(x) \in X_i$ and $z \in X_i$. So

$$F(x) = f(x) = f(z) = F(z).$$

Since f is an injection, $x = z$.

Therefore F is an injection. Hence,

$$F : X \rightarrowtail Y$$

and

$$|X| = |Y|.$$

$\square$

THEOREM 6.4. $\mathbb{N}$ *is an infinite set.*

DISCUSSION. *We show that any function with domain* $\ulcorner n \urcorner$*, for* $n \in \mathbb{N}$*, fails to be a surjection. Therefore* $\mathbb{N}$ *is not finite.*

PROOF. Assume $n \in \mathbb{N}$, and

$$f : \ulcorner n \urcorner \longrightarrow \mathbb{N}.$$

Let

$$a = 1 + \sum_{i=0}^{n-1} f(i) \in \mathbb{N}.$$

Clearly $a \notin f[\ulcorner n \urcorner]$, so f is not a surjection. Consequently, there is no $n \in \mathbb{N}$ that can be mapped surjectively onto $\mathbb{N}$. Therefore $\mathbb{N}$ is not finite. $\qquad\square$

Not only is $\mathbb{N}$ an infinite set, it is in some sense the "smallest" infinite set.

THEOREM 6.5. *If X is infinite, then $\mathbb{N} \preceq X$.*

DISCUSSION. *We shall define an injection $f : \mathbb{N} \to X$ inductively, building it up one step at a time.*

PROOF. As X is infinite, it is nonempty so must contain some element x_0. Define $f(0) = x_0$.

Now, suppose that $x_0 = f(0), x_1 = f(1), \ldots, x_n = f(n)$ have all been chosen, so that

$$ f|_{\{0,1,\ldots,n\}} = f|_{\ulcorner n+1 \urcorner} : k \mapsto x_k $$

is injective. As X is infinite, the function $f|_{\ulcorner n+1 \urcorner}$ that we have defined cannot be surjective. So there exists some x_{n+1} in $X \setminus \{x_0, \ldots, x_n\}$. Define $f(n+1) = x_{n+1}$. Continuing in this way, we attain an injection f defined on all of $\mathbb{N}$. $\qquad\square$

REMARK. The astute reader may have noticed that in the previous proof, we end up making an infinite number of choices of elements of X.

DEFINITION. Cardinality, $\aleph_0$ We use the expression $\aleph_0$ (read "aleph nought"[1]) for the size of $\mathbb{N}$. That is

$$ \aleph_0 := |\mathbb{N}|. $$

The size of a set is called the cardinality of the set. Any set that is bijective with $\mathbb{N}$ has cardinality $\aleph_0$. A finite set has cardinality equal to the unique natural number with which it is bijective.

---

[1]$\aleph$ is the first letter of the Hebrew alphabet.

DEFINITION. ▢Countable▢ A set that is finite or has cardinality $\aleph_0$ is called a countable set.

We are not formally developing the idea of cardinality. This would require working with ordinals, which would distract us from more immediate mathematical interests. However, we shall use the language and conventions of cardinals where it is intuitive and does not interfere with our program.

6.3 Uncountable Sets

In this section we prove one of the most remarkable results of modern mathematics. There are sets that are not countable. When this result was first communicated in 1878 by Georg Cantor, it astonished the mathematical world. It follows from this result that, in a most meaningful sense, there are different sizes of infinity. Suppose X is not a countable set. By Theorem 6.5, $\mathbb{N} \preceq X$. By the Schröder-Bernstein theorem, if $X \preceq \mathbb{N}$, then $|\mathbb{N}| = |X|$, and X would be countable. So if X is not countable, $X \not\preceq \mathbb{N}$, and $\mathbb{N} \prec X$. That is,

$$\aleph_0 < |X|.$$

DEFINITION. ▢Uncountable▢ A set that is not countable is called uncountable.

Of course, we have yet to show that there are any uncountable sets.

NOTATION. ▢$\prec$▢ Let X and Y be sets. Then $X \prec Y$ provided that

$$X \preceq Y$$

and

$$|X| \neq |Y|.$$

We write $|X| \leq |Y|$ if $X \preceq Y$, and $|X| < |Y|$ if $X \prec Y$.

DEFINITION. Power set, $P(X)$ Let X be a set. Then

$$P(X) \ = \ \{Y \mid Y \subseteq X\}.$$

$P(X)$ is called the power set of X. It is the set of all subsets of X.

The next theorem, due to Cantor, is one of the most remarkable results of mathematics. It not only proves the existence of an uncountable set, it implies that the power set necessarily generates sets of larger cardinality and thereby provides a means of constructing infinitely many different infinite cardinalities.

THEOREM 6.6. *Let X be a set. Then*

$$\mid X \mid < \mid P(X) \mid.$$

DISCUSSION. *To prove this result we need to show that a bijection between a set and its power set is impossible. How do we show the impossibility of such a function? We can assume that such a bijection exists and derive a contradiction. Alternatively, we can show that any function from a set to its power set necessarily fails to be a surjection— which is nearly the same thing, and more elegant. We need to show that any function from a set to its power set "misses" some elements of the power set. We shall use a technique known as a diagonal argument to construct an element of the power set that is not in the range of the function. The domain, X, acts like an index to keep track of the elements of the range of the function (this is another use for functions). We construct an element $Y \in P(X)$ not in the range of the function by adding $x \in X$ to Y iff x is not in the element of $P(X)$ indexed by x (that is, x is not in the image of x under the function). It is easy to show that this subset Y of X is not in the range of the function, and the function therefore fails to be a surjection onto $P(X)$.*

PROOF. We observe that the function $g : X \to P(X)$ defined by

$$g(x) \ = \ \{x\}$$

is an injection. In the case that $X = \emptyset$, g is the empty function—that is, the function whose graph is the $\emptyset$. (You should check that the empty function is an injection.) Therefore

$$| X | \leq | P(X) | .$$

Let

$$f : X \to P(X).$$

We define

$$Y := \{x \in X \mid x \notin f(x)\}.$$

DISCUSSION. *Recall that for every $x \in X$, $f(x)$ is a subset of X. Thus it makes sense to consider whether x is an element of $f(x)$. The self-referential flavor of this argument makes it challenging on the first reading!*

Clearly,

$$Y \subseteq X.$$

Is $Y \in f[X]$? Suppose it were, so $Y = f(x_0)$ for some x_0 in X. But then, x_0 would be in Y iff x_0 were not in $f(x_0) = Y$. This is impossible, contradicting the assumption that Y is in $f[X]$.   □

You might try to repair f by modifying it to include in its range the diagonal set we constructed. Applying the diagonal argument again will identify a new element missing from the range of the modified function. In fact, most elements of the codomain are missing from the range of the function, although this is not immediately obvious from the proof.

You still might be confused by why this is called a diagonal argument. This will be obvious when we apply the technique to infinite binary sequences.

If X is finite the theorem is obvious. Indeed, if there is $n \in \mathbb{N}$ such that $| X | = n$, then $| P(X) | = 2^n$. Theorem 6.6 implies that any set, including an infinite set, is strictly smaller than its power set. In fact, by iterating the applications of the power set function to $\mathbb{N}$, it is easily seen that there are infinitely many infinite sets of distinct cardinality in the sequence of sets

$$\langle \mathbb{N}, P(\mathbb{N}), P(P(\mathbb{N})), \ldots \rangle.$$

(What is the cardinality of the union over this sequence? What is the cardinality of the power set of that union?)

We shall prove that two more sets are uncountable. Both of these are sets of mathematical interest. We shall show first that infinite binary sequences are uncountable. Infinite binary sequences are functions from $\mathbb{N}$ to $\ulcorner 2 \urcorner$. As we shall see, there is a very close relationship between infinite binary sequences and the power set of $\mathbb{N}$. More generally, the collection of all functions from one set to another can be of mathematical interest. We introduce a notation for such collections.

NOTATION. Y^X Let X and Y be sets. The set of all functions with domain X and codomain Y is written Y^X.

Do not confuse this with exponentiation. However, if X and Y are finite,

$$\mid Y^X \mid \; = \; \mid Y \mid^{|X|}.$$

The set of all functions from some set X into $\ulcorner 2 \urcorner$ is in bijective correspondence with $P(X)$.

PROPOSITION 6.7. *Let X be a set, and define $F : \ulcorner 2 \urcorner^X \to P(X)$ by: for $\chi \in \ulcorner 2 \urcorner^X$,*

$$F(\chi) \; = \; \chi^{-1}(1).$$

That is $F(\chi) \; = \; \{x \in X \mid \chi(x) \; = \; 1\}$. Then $F : \ulcorner 2 \urcorner^X \to P(X)$ is a bijection.

PROOF. The proof is left as an exercise.   □

The existence of this bijection allows us to easily prove the following theorem.

THEOREM 6.8. *The set of infinite binary sequences is bijective with* $P(\mathbb{N})$ *and is therefore uncountable.*

PROOF. By Proposition 6.7

$$ |\ulcorner 2 \urcorner^{\mathbb{N}}| = |P(\mathbb{N})|. $$

By Theorem 6.6,

$$ |\mathbb{N}| \prec |P(\mathbb{N})|. $$

Therefore $\ulcorner 2 \urcorner^{\mathbb{N}}$ is uncountable. □

NOTATION. $2^{\aleph_0}$ We use $2^{\aleph_0}$ for the cardinality of $\ulcorner 2 \urcorner^{\mathbb{N}}$.

It is worth illustrating by an application to infinite binary sequences why the technique used to prove Theorem 6.6 is called a diagonal argument (sometimes called the second diagonal argument to distinguish it from the "first diagonal argument" in Section 6.4). Let $f : \mathbb{N} \to \ulcorner 2 \urcorner^{\mathbb{N}}$. We prove that f is not a surjection by direct application of the diagonal argument in the proof of Theorem 6.6. We enumerate all the elements in the range of f; each one is a sequence of 0s and 1s.

$$
\begin{array}{rcllllll}
f(0) & = & a_{00} & a_{01} & a_{02} & a_{03} & \dots & a_{0i} & \dots \\
f(1) & = & a_{10} & a_{11} & a_{12} & a_{13} & \dots & a_{1i} & \dots \\
f(2) & = & a_{20} & a_{21} & a_{22} & a_{23} & \dots & a_{2i} & \dots \\
f(3) & = & a_{30} & a_{31} & a_{32} & a_{33} & \dots & a_{3i} & \dots
\end{array}
$$

We now create a sequence by altering the diagonal elements of this infinite array. Let s be the sequence of diagonal elements

$$ \langle 1 - a_{00}, 1 - a_{11}, \dots, 1 - a_{ii}, \dots \rangle. $$

The sequence s is an element of $\ulcorner 2 \urcorner^{\mathbb{N}}$, and differs from every element in the range of f: indeed, s differs from $f(i)$ in at least the ith slot.

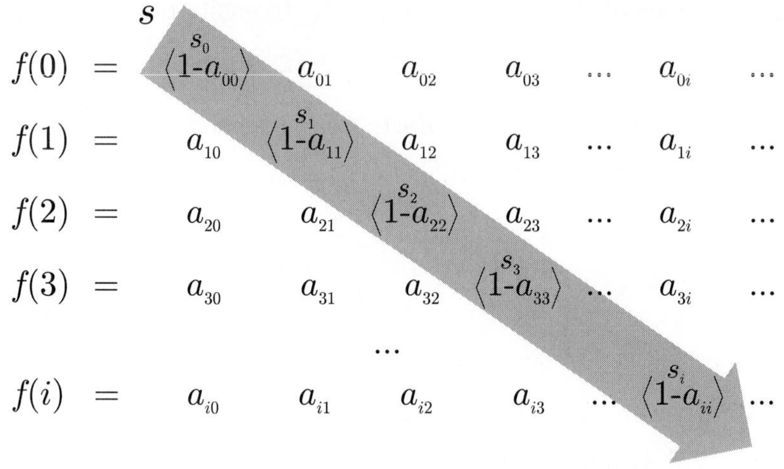

$$f(0) = \left\langle \overset{s_0}{1\text{-}a_{00}} \right\rangle \quad a_{01} \quad a_{02} \quad a_{03} \quad \dots \quad a_{0i} \quad \dots$$

$$f(1) = a_{10} \quad \left\langle \overset{s_1}{1\text{-}a_{11}} \right\rangle \quad a_{12} \quad a_{13} \quad \dots \quad a_{1i} \quad \dots$$

$$f(2) = a_{20} \quad a_{21} \quad \left\langle \overset{s_2}{1\text{-}a_{22}} \right\rangle \quad a_{23} \quad \dots \quad a_{2i} \quad \dots$$

$$f(3) = a_{30} \quad a_{31} \quad a_{32} \quad \left\langle \overset{s_3}{1\text{-}a_{33}} \right\rangle \quad \dots \quad a_{3i} \quad \dots$$

$$\dots$$

$$f(i) = a_{i0} \quad a_{i1} \quad a_{i2} \quad a_{i3} \quad \dots \quad \left\langle \overset{s_i}{1\text{-}a_{ii}} \right\rangle \dots$$

FIGURE 6.2 The second diagonal argument

Hence,

$$s \notin f[\mathbb{N}],$$

and so f is not a surjection. We leave it as an exercise to show that s is the diagonal set Y constructed in the proof of Theorem 6.6, where $X = \mathbb{N}$. More precisely, s is the image of Y under the natural bijection from $P(\mathbb{N})$ to $\ulcorner 2 \urcorner^{\mathbb{N}}$ of Proposition 6.7. □

We consider another set of mathematical interest, the set of all infinite decimal sequences. This set has a close relationship with the closed interval $[0, 1]$. Understanding this relationship requires a deeper, more formal understanding of the real numbers than most students have been exposed to in calculus, and we postpone the detailed discussion of this relationship until Section 8.9. With some modifications, the following theorem will prove that $[0, 1]$ is uncountable, and therefore $\mathbb{R}$ is uncountable (see Section 8.9).

THEOREM 6.9. *The set of infinite decimal expansions is uncountable. In fact,*

$$| \ulcorner 10 \urcorner^{\mathbb{N}} | = 2^{\aleph_0}.$$

DISCUSSION. *The identity function on the infinite binary sequences into the infinite decimal sequences is clearly an injection. We shall construct an injection from the infinite decimal sequences to infinite binary sequences. The theorem will follow from the Schröder-Bernstein theorem.*

PROOF. It is obvious that

$$|\ulcorner 2 \urcorner^{\mathbb{N}}| \leq |\ulcorner 10 \urcorner^{\mathbb{N}}| \qquad (6.10)$$

(why?). We shall define an injection

$$f : \ulcorner 10 \urcorner^{\mathbb{N}} \to \ulcorner 2 \urcorner^{\mathbb{N}}.$$

Let $x \in \ulcorner 10 \urcorner^{\mathbb{N}}$. So

$$x = \langle x_j \mid j \in \mathbb{N} \rangle$$

where x_j is the jth member of the sequence x and

$$0 \leq x_j \leq 9.$$

We want to define a binary sequence $s(x)$ that "encodes" x. There are many ways to do it. One is to look at blocks of 10 bits (short for "binary digits"), and, in the jth such block, have 9 of the bits 0 and put a 1 in the $x_j{}^{\text{th}}$ slot. Formally, given an infinite decimal sequence x, we define a binary sequence

$$f(x) = \langle y_i \mid i \in \mathbb{N} \rangle$$

so that $y_i = 1$ if there is $j \in \mathbb{N}$ such that

$$i = 10j + x_j.$$

Otherwise $y_i = 0$. We thereby define a function

$$f : \ulcorner 10 \urcorner \to \ulcorner 2 \urcorner.$$

We show that f is an injection. Let x and y be distinct elements of $\ulcorner 10 \urcorner^{\mathbb{N}}$. Then there is some $j \in \mathbb{N}$ such that

$$x_j \neq y_j.$$

Then

$$10j + x_j \neq 10j + y_j.$$

Let $i = 10j + x_j$. Then $f(x)$ and $f(y)$ differ in the ith component. That is,

$$(f(x))_i = 1 \neq 0 = (f(y))_i.$$

Therefore f is an injection and

$$|\ulcorner 10 \urcorner^{\mathbb{N}}| \preceq |\ulcorner 2 \urcorner^{\mathbb{N}}|. \tag{6.11}$$

By the Schröder-Bernstein theorem, (6.10) and (6.11) yield

$$|\ulcorner 10 \urcorner^{\mathbb{N}}| = 2^{\aleph_0}.$$

$\square$

We prove in Section 8.9 that

$$|\,[0,1]\,| = |\ulcorner 10 \urcorner^{\mathbb{N}}|,$$

essentially by identifying a real number with its decimal expansion. If we assume this result, we can easily prove that the real numbers are uncountable.

COROLLARY 6.12. $\mathbb{R}$ *is uncountable.*

6.4 Countable Sets

The uncountable sets we have identified so far have a certain structural characteristic in common. We have shown that the set of all functions from a fixed infinite domain to a fixed codomain of at least two elements is uncountable. Cantor's theorem that the power set of an infinite countable set is uncountable can be interpreted this way as well. If X is a set, then $P(X)$ can be understood as $\ulcorner 2 \urcorner^X$, the set of all functions from X to $\ulcorner 2 \urcorner$. In the case of finite sets, X and Y, the set of all functions from X to Y, Y^X, has cardinality $|\,Y\,|^{|X|}$. That is, the cardinality of

$$\{f \subseteq X \times Y \mid f \text{ is a function}\}$$

is an exponential function of $\mid X \mid$. Of course, exponential functions grow relatively quickly. For finite sets, the cardinality of the union of disjoint sets is the sum of the cardinalities of the sets. The cardinality of the direct product of two finite sets is the product of the cardinalities. What happens to the union or the direct product of countable infinite sets? Can the set operations of union and direct product generate uncountable sets from countable sets? We answer the questions for unions (addition) first.

The following proposition will simplify some of the technical details in the arguments that follow.

PROPOSITION 6.13. *Let X and Y be sets. Then there is a surjection $f : X \to Y$ iff $\mid Y \mid \leq \mid X \mid$.*

DISCUSSION. *We shall use the level sets of the surjection f to define the injection from Y to X. This uses the machinery of equivalence relations developed in Chapter 2 with the Axiom of Choice.*

PROOF. ($\Rightarrow$)
Let X, Y, and f be as in the statement of the proposition. Let

$$\widehat{f} : X/f \to Y$$

be the canonical bijection associated with f that was defined in Section 2.4. We ask whether there is an injection $g : X/f \to X$ where $g([x]) \in [x]$. Recall that X/f is the collection of level subsets of X, with respect to f, and is a partition of X. Why not simply choose an element from each equivalence class and define g to be the function from X/f to X defined by these choices?

DISCUSSION. *The Axiom of Choice is the assertion that such "choice" functions exist.*

The function g is clearly an injection, so

$$g \circ \widehat{f}^{-1} : Y \to X$$

is an injection. Therefore if there is a surjection $f : X \to Y$, then $|Y| \le |X|$.

$(\Leftarrow)$

The proof of this implication is left as an exercise. □

THEOREM 6.14. *Cantor's theorem* *Let* $\{X_n \mid n \in \mathbb{N}\}$ *be a family of sets such that* X_n *is countable for all* $n \in \mathbb{N}$, *and* $X = \bigcup_{n \in \mathbb{N}} X_n$. *Then*

$$|X| \le \aleph_0.$$

DISCUSSION. *This theorem, also due to Cantor, is the key result for proving that sets are countable. It is proved by a technique also called a diagonal argument (sometimes called the first diagonal argument). We use the index set* $\mathbb{N}$ *to construct an infinite array and use that array to illustrate an enumeration of the union. This enumeration is a surjection from* $\mathbb{N}$ *to* X.

PROOF. For $n \in \mathbb{N}$, X_n is countable and by Proposition 6.13 there is a surjection

$$f_n : \mathbb{N} \to X_n.$$

Use the functions f_n to construct an infinite array. The 0th column will contain all the elements of X_0, in the order $f_0(0), f_0(1), f_0(2), \ldots$. (It does not matter if the same element is listed multiple times.) The next column has the elements of X_1 in the order $f_1(0), f_1(1), f_1(2)$, and so on. We define a function $g : \mathbb{N} \to X$ by traversing this array along the northeast to southwest diagonals, viz. $g(0) = f_0(0), g(1) = f_1(0), g(2) = f_0(1), g(3) = f_2(0), g(4) = f_1(1), g(5) = f_0(2), g(6) = f_3(0)$, and so on.

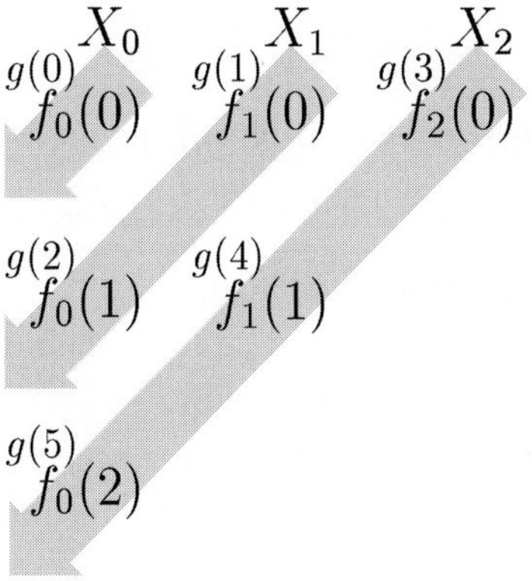

$$X_0 \qquad X_1 \qquad X_2$$

$$g(0) \qquad g(1) \qquad g(3)$$
$$f_0(0) \qquad f_1(0) \qquad f_2(0)$$

$$g(2) \qquad g(4)$$
$$f_0(1) \qquad f_1(1)$$

$$g(5)$$
$$f_0(2)$$

FIGURE 6.3 The first diagonal argument

Then g is a surjection because every element of $\bigcup X_n$ occurs in the array and is therefore in the range of g. By Proposition 6.13,

$$|X| \leq \aleph_0.$$

$\square$

COROLLARY 6.15. *Let A be a countable set and $\{X_\alpha \mid \alpha \in A\}$ be a family of countable sets indexed by A. Then*

$$\left| \bigcup_{\alpha \in A} X_\alpha \right| \leq \aleph_0.$$

PROOF. Since A is countable, there is a surjection

$$f : \mathbb{N} \to A.$$

Therefore

$$\bigcup_{\alpha \in A} X_\alpha = \bigcup_{n \in \mathbb{N}} X_{f(n)}.$$

By Cantor's theorem 6.14,

$$\left| \bigcup_{\alpha \in A} X_\alpha \right| \leq \aleph_0.$$

$\square$

COROLLARY 6.16. $\mathbb{Z}$ *is countable.*

DISCUSSION. *Without too much effort, we could define a bijection from $\mathbb{N}$ to $\mathbb{Z}$. Instead we shall prove the existence of the bijection without explicitly defining a bijection.*

PROOF. Let $f : \mathbb{N} \to \mathbb{Z}$ be such that

$$f(n) \;=\; -n.$$

Then $f[\mathbb{N}]$ is countable. By Cantor's theorem

$$\mathbb{Z} \;=\; \mathbb{N} \cup f[\mathbb{N}]$$

is countable. $\qquad\qquad\qquad\qquad\qquad\qquad\qquad\qquad\qquad\qquad\qquad$ □

We turn our attention to direct products.

THEOREM 6.17. *If $n \in \mathbb{N}$, and $X_1, X_2, \ldots, X_n$ are countable sets, then*

$$X_1 \times X_2 \times \cdots X_n$$

is countable.

PROOF. We assume that all of the factors, $X_1, \ldots, X_n$ are nonempty. We argue by induction on the number of factors.

BASE CASE: $n = 2$.

$$X_1 \times X_2 \;=\; \bigcup_{x \in X_2} X_1 \times \{x\}.$$

For each $x \in X_2$,

$$|\, X_1 \,| \;=\; |\, X_1 \times \{x\} \,|\,.$$

By Corollary 6.15, $X_1 \times X_2$ is countable.

INDUCTION STEP:

Assume that for any collection of n countable sets $X_1, \ldots X_n$, the product $X_1 \times \cdots \times X_n$ is countable. Let $X_1, \ldots, X_{n+1}$ be countable nonempty sets. Then

$$X_1 \times \cdots \times X_{n+1} \;=\; (X_1 \times \cdots \times X_n) \times X_{n+1}.$$

By the induction hypothesis, $X_1 \times \cdots \times X_n$ is countable, and by the base case, the direct product of two countable sets is countable. Therefore, $X_1 \times \cdots \times X_{n+1}$ is countable.   □

COROLLARY 6.18. $\mathbb{Q}$ *is countable.*

PROOF. Let $f : \mathbb{Z} \times \mathbb{Z} \to \mathbb{Q}$ be defined by

$$f(a, b) = \begin{cases} a/b & \text{if} \quad b \neq 0 \\ 0 & \text{otherwise.} \end{cases}$$

Then f is a surjection, and by Proposition 6.13, $\mathbb{Q}$ is countable.   □

We have evaluated the nested sequence of sets,

$$\mathbb{N} \subsetneq \mathbb{Z} \subsetneq \mathbb{Q} \subsetneq \mathbb{R}.$$

These are important mathematical sets and, with the exception of $\mathbb{R}$, they are countable. We investigate the cardinality of one more set between $\mathbb{Q}$ and $\mathbb{R}$.

DEFINITION. Algebraic real number, $\mathbb{K}$ A real number α is algebraic if there is a polynomial p (not identically 0) with integer coefficients such that $p(\alpha) \;=\; 0$. We shall denote the set of all algebraic numbers by $\mathbb{K}$.

Any rational number $a/b \in \mathbb{Q}$ is algebraic since a/b is a root of the polynomial

$$p(x) \;=\; bx - a.$$

Moreover, in Example 3.23, we showed that $\sqrt{2}$ is irrational, and it is clearly algebraic since it is a root of $x^2 - 2$. Therefore we have

$$\mathbb{N} \subsetneq \mathbb{Z} \subsetneq \mathbb{Q} \subsetneq \mathbb{K} \subseteq \mathbb{R}.$$

Finally we prove that $\mathbb{K} \neq \mathbb{R}$ by showing that $\mathbb{K}$ is countable.

THEOREM 6.19. $\mathbb{K}$ *is countable.*

DISCUSSION. *This result is proved by showing that the algebraic real numbers can be constructed by a countable procedure. That is,* $\mathbb{K}$ *may be built by adding to* $\mathbb{Q}$ *countably many elements at a time countably many times. Cantor's theorem implies that any set so constructed will be countable.*

PROOF. Let $n \in \mathbb{N}$ and define $f : \prod_{i=0}^{n} \mathbb{Z} \to \mathbb{Z}[x]$ by

$$f(a_0, \dots, a_n) = \sum_{i=0}^{n} a_i x^i.$$

By Corollary 6.16, $\mathbb{Z}$ is countable. By Theorem 6.17, $\prod_{i=0}^{n} \mathbb{Z}$ is countable. The range of f is the set of polynomials with integer coefficients with degree $\leq n$ (or the polynomial identically equal to 0). By Proposition 6.13, the range of a function with a countable domain is countable as well. Therefore the set of polynomials of degree $\leq n$ is countable.

Let P_n be the set of polynomials with integer coefficients of degree $\leq n$. Then

$$\mathbb{Z}[x] = \bigcup_{i=0}^{\infty} P_n.$$

By Theorem 6.14, $\mathbb{Z}[x]$ is countable. By Theorem 4.10, if $p(x)$ is a polynomial with real coefficients of degree n, it has at most n real roots. Let

$$Z_p = \{\alpha \mid p(\alpha) = 0\}.$$

So Z_p is finite for every polynomial p. Applying Cantor's theorem (Theorem 6.14) again,

$$\mathbb{K} = \bigcup_{p \in \mathbb{Z}[x]} Z_p$$

is countable. □

COROLLARY 6.20. $\mathbb{K} \neq \mathbb{R}$

Since $\mathbb{K}$ is countable and $\mathbb{R}$ is uncountable, $\mathbb{K}$ is a proper subset of $\mathbb{R}$.

DEFINITION. Transcendental number A real number that is not algebraic is called a transcendental number.

Corollary 6.20 states that there are transcendental numbers. This is an existence claim in which no witness to the claim is produced. Rather it is an example of a counting argument (on infinite sets). There are too many real numbers for them all to be algebraic. By the end of the nineteenth century it was proved that π and e are transcendental, but these proofs are much more complicated than Cantor's existence proof above, which is, in essence, a very clever application of the pigeonhole principle.

COROLLARY 6.21. *There are uncountably many transcendental numbers.*

PROOF. Let T be the set of transcendental numbers. As

$$|\,\mathbb{R}\,| = |\,T \cup \mathbb{K}\,| > \aleph_0$$

and $\mathbb{K}$ is countable, T must be uncountable. $\square$

So we have shown that

$$\mathbb{N} \subsetneq \mathbb{Z} \subsetneq \mathbb{Q} \subsetneq \mathbb{K} \subsetneq \mathbb{R}.$$

However,

$$|\,\mathbb{N}\,| = |\,\mathbb{Z}\,| = |\,\mathbb{Q}\,| = |\,\mathbb{K}\,| < |\,\mathbb{R}\,|.$$

6.5 Functions and Computability

In Section 1.3 we made the offhand comment that most functions are not defined by rules (by which we meant instructions for computing the function). We consider a rule to be an instruction (in some language) of finite length. Functions that are unambiguously defined by a rule of finite length are called *computable,* or *recursive functions.* Naturally there is a complicated mathematical definition of recursive functions, but we shall dispense with the formalities and say that a function is recursive, or computable, if there is an instruction (of finite length) for finding the image of any element in the domain. How many computable functions are there?

We shall restrict our investigation to functions from $\mathbb{N}$ to $\mathbb{N}$. We consider functions as graphs of functions. That is, every subset of $P(\mathbb{N} \times \mathbb{N})$ that satisfies the definition of a function is a function in $\mathbb{N}^{\mathbb{N}}$. Are all such functions computable? It is obvious that

$$2^{\aleph_0} \leq |\, \mathbb{N}^{\mathbb{N}} \,|$$

(why?). In fact you can show that the sets are bijective. So there are uncountably many functions in $\mathbb{N}^{\mathbb{N}}$. How many instructions for computing functions are there? An instruction is a finite string, or sequence, of symbols. For instance, an instruction for the function that squares natural numbers is

$$f(x) \;=\; x^2.$$

This is a finite sequence of seven symbols. The instruction gives enough information to compute the image of any natural number. There are many other rules for computing this function. For instance the rule

$$f(x) \;=\; x \cdot x$$

obviously defines the same function, but the instruction is different—it contains one more symbol. Consider the set of all possible instructions for computing functions of natural numbers. How are the instructions formulated? One produces a finite sequence of symbols that forms an explicit guide for computing the image of any natural number.

Let X be the set of all symbols appearing in instructions for computing functions of natural numbers. The set X will include letters, digits, symbols for operations, symbols for relations, and potentially any other symbol that you might see in a book on mathematics. How large is X? If you require that every symbol appear in some actual dictionary, it would clearly be finite. You will probably wish to allow any natural number to appear in the instruction. However, although there are infinitely many natural numbers, we need only 10 symbols to name them all. It seems that we can reasonably require that X is finite, but as it turns out, we can allow for X to be countably infinite without changing our conclusion.

If there is *any* language with countably many symbols in which the set of all instructions for computing functions could be written, then we may assume that X is countable. If F is an instruction or rule (and hence a finite sequence of symbols from X), then there is $N \in \mathbb{N}$ such that

$$F \in X^N.$$

So it is easily seen that the set of all possible instructions for elements of $\mathbb{N}^{\mathbb{N}}$, I, satisfies

$$I \preceq \bigcup_{N \in \mathbb{N}} X^N.$$

For $N \in \mathbb{N}$, X^N is the direct product of N factors of X, and by Theorem 6.17,

$$|X^N| \leq \aleph_0.$$

The set $\bigcup_{N \in \mathbb{N}} X^N$ is the countable union of countable sets, and by Theorem 6.14 is countable. Therefore there are uncountably many functions of natural numbers that are not defined by rules.

For a more thorough treatment of set theory, see the book [5] by Yiannis Moschovakis.

6.6 Exercises

EXERCISE 6.1. Let $f : X \rightarrowtail Y$ and $g : Y \rightarrowtail Z$. Prove that

$$g \circ f : X \rightarrow Z$$

is a bijection.

EXERCISE 6.2. Prove that equinumerosity is an equivalence relation.

EXERCISE 6.3. Prove that the relation on sets $\preceq$ is reflexive and transitive.

EXERCISE 6.4. In the proof of the Schröder-Bernstein theorem, define a function

$$G(x) = \begin{cases} g^{-1}(x) & \text{if} & x \in X_i \\ f(x) & \text{if} & x \in X_e \\ g^{-1}(x) & \text{if} & X_o. \end{cases}$$

Prove that $G : X \rightarrowtail Y$.

EXERCISE 6.5. Let $n \in \mathbb{N}$. Prove that

$$|P(\ulcorner n \urcorner)| = 2^n.$$

EXERCISE 6.6. Let $X = \{0, 1, 2\}$. Write down some function $f : X \rightarrow P(X)$. For this particular f, what is the set Y of Theorem 6.6?

EXERCISE 6.7. Let X be a set and define a sequence of sets $\langle X_n \mid n \in \mathbb{N} \rangle$ by

$$X_0 = X$$

and

$$X_{n+1} = P(X_n).$$

Let $Y = \bigcup_{n=0}^{\infty} X_n$. Prove

$$(\forall n \in \mathbb{N}) \ |X_n| < |Y|.$$

EXERCISE 6.8. Let X and Y be finite sets. Prove that

$$| X^Y | = | X |^{|Y|} .$$

EXERCISE 6.9. Prove Proposition 6.7.

EXERCISE 6.10. Let $f : \mathbb{N} \to \ulcorner 2 \urcorner^{\mathbb{N}}$ and for $i, j \in N^+$

$$a_{ij} = (f(i))_j.$$

(That is, a_{ij} is the jth term of the ith sequence.) Let s be the "diagonal" sequence

$$s = \langle 1 - a_{nn} \mid n \in N^+ \rangle.$$

We know that $s \notin f[\mathbb{N}]$. If $F : \ulcorner 2 \urcorner^{\mathbb{N}} \rightarrowtail P(\mathbb{N})$ is the bijection in Proposition 6.7, then $F \circ f : \mathbb{N} \to P(\mathbb{N})$. Prove that $F(s)$ is the "diagonal" set of Theorem 6.6 (where $X = \mathbb{N}$, and $F \circ f$ is the enumeration of subsets of $\mathbb{N}$) and hence that $F(s) \notin (F \circ f)[\mathbb{N}]$.

EXERCISE 6.11. Prove that if $X \subseteq Y$ and X is uncountable, then Y is uncountable.

EXERCISE 6.12. Let X be an uncountable set, Y be a countable set, and $f : X \to Y$. Prove that some element of Y has an uncountable pre-image.

EXERCISE 6.13. Complete the proof of Proposition 6.13.

EXERCISE 6.14. Define an explicit bijection from $\mathbb{N}$ to $\mathbb{Z}$.

EXERCISE 6.15. Prove that $| \mathbb{K} \setminus \mathbb{Q} | = \aleph_0$.

EXERCISE 6.16. Prove that

$$e = \sum_{n=0}^{\infty} \frac{1}{n!}$$

is irrational. (Hint: Argue by contradiction. Assume $e = \frac{p}{q}$ and multiply both sides by $q!$. Rearrange the equation to get an integer equal to an infinite sum of rational numbers that converges to a number in the open interval $(0, 1)$.)

Remark: This was also Exercise 3.32. Is it easier now?

EXERCISE 6.17. Suppose that $a, b, c, d \in \mathbb{R}$, $a < b$, and $c < d$. Prove

(i) The open interval (a, b) is bijective with the open interval (c, d).

(ii) The closed interval $[a, b]$ is bijective with the closed interval $[c, d]$.

(iii) The open interval $(0, 1)$ is bijective with the closed interval $[0, 1]$.

(iv) The open interval (a, b) is bijective with the closed interval $[c, d]$.

(v) $| [0, 1] | = | \mathbb{R} |$.

EXERCISE 6.18. Construct explicit bijections for each of the pairs of sets in Exercise 6.17.

EXERCISE 6.19. Let $f(x)$ be a nonzero polynomial with integer coefficients, and suppose $\alpha \in \mathbb{R}$ is transcendental. Prove that $f(\alpha)$ is transcendental.

EXERCISE 6.20. Let $F : \mathbb{K} \to \mathbb{R}$ be defined by the following: If $x \in \mathbb{K}$, $F(x)$ is the lowest degree of a polynomial with integer coefficients for which x is a root. Is F well-defined?

EXERCISE 6.21. Let $a \in \mathbb{R}$ be a root of a polynomial with rational coefficients. Prove that a is a root of a polynomial with integer coefficients and is therefore an algebraic number.

EXERCISE 6.22. For each of the following sets, state and prove whether it is bijective with $\mathbb{N}$, $P(\mathbb{N})$ or is larger than $P(\mathbb{N})$ (with respect to the relation $\prec$):

(i) the set of finite subsets of $\mathbb{N}$

(ii) the set of all permutations of finite sets of natural numbers

(iii) the set of finite sequences of natural numbers

(iv) the set of finite sequences of integers

(v) the set of finite sequences of algebraic numbers

(vi) the set of finite sequences of real numbers

(vii) the set of infinite sequences of natural numbers

(viii) the set of infinite sequences of real numbers

(ix) countable subsets of $\mathbb{R}$

(x) $\mathbb{N}^{\mathbb{R}}$

(xi) $\mathbb{R}^{\mathbb{R}}$

You may use the fact that $|\mathbb{R}| = 2^{\aleph_0}$.

EXERCISE 6.23. Prove that $|\mathbb{R}^{\mathbb{R}}| \geq |P(\mathbb{R})|$.

CHAPTER 7

Divisibility

In this chapter we investigate divisibility. It may seem peculiar that we would investigate a topic that you have studied since elementary school, but do not be fooled by the apparent simplicity of the subject. The study of divisibility of integers is part of number theory. Geometry and number theory are the oldest areas of mathematical study, and they are still active fields of mathematical research.

7.1 Fundamental Theorem of Arithmetic

DEFINITION. Divides, factor Let $a, b \in \mathbb{Z}$. We say that a divides b, or a is a factor of b, if

$$(\exists c \in \mathbb{Z}) \; a \cdot c \; = \; b.$$

We write this as $a \mid b$. If a does not divide b, we write $a \nmid b$.

Divisibility is the central idea of number theory. It is precisely because one integer need not be a factor of another integer, or a pair of integers may fail to have nontrivial common factors, that divisibility provides insight into the structure of integers. Put another way, consider the definition of divisibility applied to rational numbers—you will find that it does not provide any insight at all since a nonzero rational number is a factor of any other rational number. Furthermore, many of the properties of integers with regard to divisibility generalize to other interesting mathematical structures. You will see an example of this in Section 7.5.

DEFINITION. Prime number Let $p \in \mathbb{N}$. We say that p is a prime number if $p > 1$ and the only positive factors of p are p and 1.

DEFINITION. Relatively prime Let $a, b \in \mathbb{Z}$. We say that a and b are relatively prime if they have no common factor greater than 1.

DEFINITION. Integer combination Let $a, b, c \in \mathbb{Z}$. Then c is an integer combination of a and b if

$$(\exists m, n \in \mathbb{Z}) \ c \ = \ ma + nb.$$

PROPOSITION 7.1. *Let $a, b \in \mathbb{Z}$. If a and b are relatively prime, then $a - b$ and b are relatively prime.*

DISCUSSION. *We shall prove the contrapositive by showing that any common factor of $a - b$ and b is also a factor of a.*

PROOF. Let $c > 1$ be a common factor of a and $a - b$. So

$$(\exists m \in \mathbb{Z}) \ b \ = \ cm$$

and

$$(\exists n \in \mathbb{Z}) \ a - b \ = \ cn.$$

Then

$$c(m + n) \ = \ a$$

and so $c \mid a$. Therefore if a and b are relatively prime, then $a - b$ and b are relatively prime. $\square$

PROPOSITION 7.2. *Let a and b be integers. If a and b are relatively prime, then*

$$(\exists m, n \in \mathbb{Z}) \ ma + nb \ = \ 1.$$

DISCUSSION. *We shall argue for the case in which a and b are natural numbers. Given the proposition for all pairs of relatively prime natural numbers, we may easily extend it to arbitrary pairs of relatively prime integers by changing the sign of m or n in the integer* combination. *This assumption allows us to argue by induction on the sum of the integers. The base case for this argument by induction will be $a + b = 3$. If $a = 0 = b$, then a and b are not relatively prime. If $a + b = 1$, then a and b are relatively prime and the choice of m and n is obvious. If $a = b = 1$ then a and b are relatively prime and again the choice of m and n is obvious.*

PROOF. We may assume that $a > b > 0$. We argue by induction on $a + b$.

BASE CASE: $a + b = 3$.
Then $a = 2$ and $b = 1$. So

$$a - b = 1.$$

INDUCTION STEP: Assume that the result holds for all pairs of relatively prime natural numbers with sum less than $a + b$.
By Proposition 7.1, b and $a - b$ are relatively prime. By the induction hypothesis, there are $i, j \in \mathbb{Z}$ such that

$$i(a - b) + jb = 1.$$

DISCUSSION. *If $a - b = b$, we are not in the case where we have two distinct positive numbers. How do we handle this possibility?*

Let $m = i$ and $n = j - i$. Then

$$ma + nb = 1.$$

By the induction principle, the result holds for all relatively prime pairs of natural numbers.   □

DEFINITION.   Greatest common divisor, gcd(a, b) Let $a, b \in \mathbb{Z}$. The greatest common divisor of a and b, written gcd(a, b), is the largest integer that divides both a and b.

So a and b are relatively prime iff gcd$(a, b) = 1$.

PROPOSITION 7.3. *Let $a, b, c \in \mathbb{Z}$, and assume that* $\gcd(a, b) = 1$. *If $a \mid cb$, then $a \mid c$.*

PROOF. By Proposition 7.2 there are $m, n \in \mathbb{Z}$ such that

$$ma + nb = 1.$$

Therefore

$$cma + cnb = c.$$

Clearly $a \mid cnb$ (since $a \mid cb$) and $a \mid cma$. So

$$a \mid (cma + cnb)$$

and therefore $a \mid c$.   □

PROPOSITION 7.4. *Let $a, b, c \in \mathbb{Z}$. If* $\gcd(a, b) = 1$, *$a \mid c$ and $b \mid c$, then*

$$ab \mid c.$$

PROOF. Let $m, n \in \mathbb{Z}$ be such that $am = c$ and $bn = c$. Then

$$a \mid bn.$$

By Proposition 7.3, $a \mid n$. Hence there is $k \in \mathbb{Z}$ such that

$$ak = n.$$

Therefore

$$akb = c$$

and

$$ab \mid c.$$

                                                                            □

LEMMA 7.5. *Assume*

(1) *$p \in \mathbb{N}$ is prime*

(2) *$N \geq 1$, and $a_1, \ldots, a_N \in \mathbb{Z}$*

(3) *$p \mid (\prod_{n=1}^{N} a_n)$.*

Then there is some $n \leq N$ such that $p \mid a_n$.

PROOF. Let p be a prime number. We argue by induction on N.
BASE CASE: $N = 1$
The base case is obvious.
INDUCTION STEP: Let $N > 1$ and assume that the result holds for all products of fewer than N factors.
Let

$$a = \prod_{n=1}^{N-1} a_n$$

and suppose that

$$p \mid \left(\prod_{n=1}^{N} a_n \right).$$

Then

$$p \mid a \cdot a_N.$$

If $p \mid a$, then by the induction hypothesis,

$$(\exists n < N)\ p \mid a_n.$$

Assume that p is not a factor of a; since p is prime, $\gcd(p, a) = 1$. By Proposition 7.3, $p \mid a_N$. □

THEOREM 7.6. *Fundamental Theorem of Arithmetic Let N be a natural number greater than 1. Then N may be uniquely expressed as the product of prime numbers (up to the order of the factors).*

DISCUSSION. *We permit a "product" with only one factor. So any prime number is its own unique prime factoring.*

PROOF. We argue by induction on the natural numbers greater than 1.
BASE CASE: $(N = 2)$
By the discussion preceding the proof, 2 is its own prime factoring.
INDUCTION STEP: Assume that the result holds for all natural numbers

greater than 1 and less than N. If N is prime, the result follows. If N is not prime, then there are $a, b \in \mathbb{N}$, $a < N$, and $b < N$, such that

$$a \cdot b = N.$$

By the induction hypothesis, a and b have unique prime factorings. The product of the factorings will be a prime factoring of N. Is the factoring unique up to order? Suppose that

$$N = \prod_{i=1}^{m} p_i = \prod_{j=1}^{n} q_j$$

where p_i is prime for $1 \leq i \leq m$ and q_j is prime for $1 \leq j \leq n$. Then

$$p_1 \mid \prod_{j=1}^{n} q_j.$$

By Lemma 7.5,

$$(\exists j \leq n)\, p_1 \mid q_j.$$

We may reorder the factors $q_1, \ldots, q_n$ so that $p_1 \mid q_1$. Both p_1 and q_1 are prime, so

$$p_1 = q_1.$$

Therefore

$$\prod_{i=2}^{m} p_i = \prod_{j=2}^{n} q_j < N.$$

By the induction hypothesis, $p_2, \ldots, p_m$ is a unique prime factoring of $\prod_{i=2}^{m} p_i$, so $m = n$ and $q_2, \ldots, q_n$ is a reordering of $p_2, \ldots, p_m$. Therefore $q_1 \cdots q_n$ is a reordering of $p_1 \cdots p_m$, and the prime factoring of N is unique. $\qquad\qquad\square$

REMARK. Why is the number 1 not defined to be a prime? After all, it has no factors other than itself or 1! The reason is because it is very useful to have uniqueness in the Fundamental Theorem of Arithmetic. If 1 were considered prime, it could be included arbitrarily often in the factoring of N.

7.2 Division Algorithm

The division algorithm, Theorem 7.13, is the result that guarantees that long division of natural numbers will terminate in a unique quotient and remainder with the remainder strictly smaller than the divisor. Long division is difficult and tedious for young students. Typically it is the most challenging computation that elementary school students are expected to master. You may have revisited the algorithm again when you learned to divide polynomials. Here the division algorithm says that the quotient and remainder are unique and the remainder is either identically 0 or has degree strictly smaller than the divisor. We frequently compare the arithmetic of integers and the arithmetic of polynomials, and it is the division algorithm that makes this comparison useful.

Let us extend the link between integer combinations and greatest common divisors. According to Proposition 7.2, a pair of integers are relatively prime if there is an integer combination of the pair that equals 1. This result generalizes to greatest common divisors other than 1.

THEOREM 7.7. *Let $a, b \in \mathbb{Z}$. The set of integer combinations of a and b equals the set of integer multiples of* $\gcd(a, b)$.

PROOF. Let $c = \gcd(a, b)$ and

$$M = \{kc \mid k \in \mathbb{Z}\}.$$

Since c is a divisor of a and b, there are $i, j \in \mathbb{Z}$ such that

$$a = ic$$

and

$$b = jc.$$

Let

$$I = \{ma + nb \mid m, n \in \mathbb{Z}\}.$$

We show first that $I \subseteq M$.

If $m, n \in \mathbb{Z}$, then

$$ma + nb = mic + njc = (mi + nj)c.$$

Hence every integer combination of a and b is a multiple of c and

$$I \subseteq M.$$

Now we show that $M \subseteq I$. Let $kc \in M$ and

$$r = \gcd(i, j).$$

Then there are $m, n \in \mathbb{Z}$ such that

$$rmc = ic = a \tag{7.8}$$

and

$$rnc = jc = b. \tag{7.9}$$

So $rc \mid a$ and $rc \mid b$. Hence,

$$\gcd(a, b) \geq rc \geq c.$$

However, $\gcd(a, b) = c$, and thus $r = 1$. Therefore i and j are relatively prime.

By Proposition 7.2, there is an integer combination of i and j that equals 1. Let $u, v \in \mathbb{Z}$ be such that

$$ui + vj = 1.$$

Then

$$c(ui + vj) = c$$

and

$$kc = kc(ui + vj) = k(ua + vb)$$

by equations 7.8 and 7.9. Hence,

$$kc \in I,$$

and, as k was arbitrary,

$$M \subseteq I.$$

☐

COROLLARY 7.10. *Let $a, b \in \mathbb{Z}$. Then $\gcd(a, b)$ is the smallest positive integer combination of a and b.*

Theorem 7.7 tells us that the integer combinations of a and b are precisely the integer multiples of $\gcd(a, b)$ (which happens to be the smallest positive integer combination of a and b). We think of $\gcd(a, b)$ as "generating" through multiplication the set of integer combinations of a and b.

PROPOSITION 7.11. *Let $a, b, k \in \mathbb{Z}$. Then*

$$\gcd(a, b) = \gcd(a - kb, b).$$

PROOF. If $c \in \mathbb{Z}$, $c \mid a$, and $c \mid b$, then $c \mid a - kb$. Therefore

$$\gcd(a, b) \leq \gcd(a - kb, b). \tag{7.12}$$

Likewise, if $c \mid a - kb$ and $c \mid b$, then $c \mid a$, so we get the reverse inequality of (7.12), so the two sides are equal. ☐

THEOREM 7.13. *Division algorithm Let $a, b \in \mathbb{Z}$, $b \neq 0$. Then there are unique $q, r \in \mathbb{Z}$ such that*

$$a = qb + r$$

where $0 \leq r < \mid b \mid$.

DISCUSSION. *In the division algorithm, a is called the dividend, b the divisor, q the quotient, and r the remainder.*

PROOF. Let $a, b \in \mathbb{Z}$ and $b \neq 0$. Define $I \subseteq \mathbb{N}$ by

$$I = \{a - kb \mid k \in \mathbb{Z}\} \cap \mathbb{N}.$$

I has a smallest element, $a - qb$, for some $q \in \mathbb{Z}$.
CLAIM: $0 \leq a - qb < \mid b \mid$.
PROOF OF CLAIM. We argue by cases.

CASE 1: $b > 0$

If $a - qb \geq b$, then

$$a - (q+1)b \geq 0.$$

Hence,

$$a - (q+1)b \in I.$$

However, $a - qb$ is minimal in I, so this is impossible. Therefore,

$$a - qb < \mid b \mid .$$

CASE 2: $b < 0$

If $a - qb \geq \mid b \mid$, then

$$a - qb > a - (q-1)b \geq 0.$$

As in the first case

$$a - (q-1)b \in I.$$

This is impossible since by assumption $a - qb$ is minimal in I. Therefore,

$$a - qb < \mid b \mid .$$

$$\triangleleft$$

Thus if

$$r := a - qb,$$

we have $a = qb + r$ and $0 \leq r < \mid b \mid$. It remains to show that the quotient and remainder are unique. Suppose

$$a = mb + r = nb + s$$

where $0 \leq r, s < |b|$. If $r = s$, then $mb = nb$ and $m = n$. So we assume that $r \neq s$. Without loss of generality we assume that $r < s$. Then

$$0 \leq s - r = (m-n)b < \mid b \mid .$$

So $m - n = 0$ and $r = s$, a contradiction.   □

Of course, q and r could be found by long division—that is, one may subtract multiples of b until the remainder is less than $\mid b \mid$.

7.3 Euclidean Algorithm

How do we find $\gcd(a, b)$, for $a, b \in \mathbb{N}$? One might invoke the Fundamental Theorem of Arithmetic and compare the prime decompositions of a and b. Suppose

$$a = \prod_{n=1}^{N} p_n^{r_n}$$

and

$$b = \prod_{n=1}^{N} p_n^{s_n}$$

where $r_n, s_n \in \mathbb{N}$ for $1 \leq n \leq N$. If $t_n = \min(r_n, s_n)$ for $1 \leq n \leq N$, then

$$\gcd(a, b) = \prod_{n=1}^{N} p_n^{t_n}.$$

However, finding the prime decomposition of an integer can be quite difficult. We shall define an operation on pairs of integers that after a reasonable number of applications will yield the greatest common divisor of the integers.

If $a, b \in \mathbb{N}$, $a > b > 0$, define $E : \mathbb{N}^2 \to \mathbb{N}^2$ by

$$E(a, b) = (b, r)$$

where r is the unique remainder (when dividing a by b) whose existence was proved in the division algorithm. That is, if

$$a = qb + r$$

with $0 \leq r < b$, then define

$$E(a, b) := (b, r).$$

If $b = 0$, then let

$$E(a, 0) = (a, 0).$$

Let $(a,b) \in \mathbb{N}^2$, $a > b > 0$. We define a sequence of elements in $\mathbb{N}^2$, $\langle E_i(a,b) \mid i \in \mathbb{N} \rangle$, by recursion:

$$E_0(a,b) \;=\; (a,b)$$

and if $n > 0$

$$E_n(a,b) \;=\; E(E_{n-1}(a,b)).$$

So long as $E_n(a,b)$ has nonzero components, the sequence of second components is strictly decreasing, so it is clear that the sequence must eventually become fixed on an ordered pair (see Exercise 4.11). By the division algorithm, this will occur when the second component equals 0. Let k be the smallest integer such that

$$E_k(a,b) \;=\; E_{k+1}(a,b).$$

Then we say that $\langle E_n(a,b) \mid n \in \mathbb{N} \rangle$ stabilizes at step k. For $n \geq k$,

$$E_n(a,b) \;=\; E_{n+1}(a,b) \;=\; E_k(a,b).$$

If $\langle E_n(a,b) \rangle$ stabilizes at step k, it is obvious that $k \leq b$. Typically, the sequence stabilizes much faster than this.

THEOREM 7.14. *Let $a, b \in \mathbb{N}$, $a > b > 0$. The nonzero component on which the sequence*

$$\langle E_n(a,b) | n \in \mathbb{N} \rangle$$

stabilizes is $\gcd(a,b)$.

PROOF. Let a be fixed. We argue by induction on the smaller of the integers, b.

BASE CASE: $b = 1$

Then for any $a > 1$,

$$E(a,1) \;=\; (1,0)$$

and the sequence $\langle E_n(a,1) \rangle$ stabilizes at step 1 with nonzero component 1.

INDUCTION STEP:

Let $b > 1$. Assume the result holds for all $c < b$—that is, for any $(a, c) \in \mathbb{R}^2$, where $c < b < a$, the nonzero component of the ordered pair at which the sequence $\langle E_n(a, c) \rangle$ stabilizes is $\gcd(a, c)$. We show that the nonzero component of the ordered pair at which the sequence $\langle E_n(a, b) \rangle$ stabilizes is $\gcd(a, b)$. If $a > b > 0$, then

$$E(a, b) = (b, a - qb)$$

where $0 \leq a - qb < b$. By the induction hypothesis, the nonzero component of the ordered pair at which the sequence $\langle E_n(b, a - qb) \mid n \in \mathbb{N} \rangle$ stabilizes is $\gcd(b, a - qb)$. By Proposition 7.11

$$\gcd(a, b) = \gcd(b, a - qb).$$

So the nonzero component of the ordered pair at which the sequence

$$\langle E_n(a, b) \mid n \in \mathbb{N} \rangle$$

stabilizes is $\gcd(a, b)$. By the induction principle, the result holds for all ordered pairs $(a, b) \in \mathbb{N}^2$ where $a > b > 0$. $\square$

An algorithm is a set of executable computational instructions. The Euclidean algorithm is the following set of instructions:

Given a pair of natural numbers, $a > b > 0$, compute the sequence $\langle E_n(a, b) \mid n \in \mathbb{N} \rangle$ until the sequence stabilizes. The nonzero component of the ordered pair on which the sequence stabilizes is $\gcd(a, b)$.

EXAMPLE 7.15. Let $a = 29712375$ and $b = 119119$. Find the $\gcd(a, b)$. We use the Euclidean algorithm. So

$$
\begin{aligned}
E_0(a, b) &= (a, b) \\
E_1(a, b) &= E(a, b) = (b, 51744) \\
E_2(a, b) &= E(b, 51744) = (51744, 4851) \\
E_3(a, b) &= E(51744, 4851) = (4851, 1078) \\
E_4(a, b) &= E(4851, 1078) = (1078, 539) \\
E_5(a, b) &= E(1078, 539) = (539, 0).
\end{aligned}
$$

Therefore $\gcd(a, b) = 539$. If you employ the Fundamental Theorem of Arithmetic, with some work you can determine that

$$29{,}712{,}375 = (3^2)(5^3)(7^4)(11)$$

and

$$119{,}119 = (7^2)(11)(13)(17).$$

So $\gcd(a, b) = (7^2)(11) = 539$.

7.4 Fermat's Little Theorem

NOTATION. $\mathbb{Z}_n^*$ Let $n \in \mathbb{N}$, $n \geq 2$. Then

$$\mathbb{Z}_n^* = \mathbb{Z}_n \setminus \{[0]\}.$$

LEMMA 7.16. *Let $a, n \in \mathbb{Z}$, $n \geq 2$, be such that $\gcd(a, n) = 1$. Define $\phi_a : \mathbb{Z}_n^* \to \mathbb{Z}_n^*$ by*

$$\phi_a([b]) = [ab].$$

Then ϕ_a is a permutation of $\mathbb{Z}_n^*$.

PROOF. We show that $[a], [2a], \ldots, [(n-1)a]$ are distinct elements of $\mathbb{Z}_n^*$. Let $0 < i \leq j < n$ and suppose that $ia \equiv ja \bmod n$. Then

$$n \mid ja - ia$$

and

$$n \mid (j - i)a.$$

We assume that $\gcd(n, a) = 1$, so by Proposition 7.3, $n \mid (j - i)$. However $0 \leq j - i < n$, so $j - i = 0$ and $i = j$. Hence, if $0 < i < j < n$,

$$[ia] \neq [ja].$$

It follows that ϕ_a is an injection from $\mathbb{Z}_n^*$ to $\mathbb{Z}_n^*$. Any injection from a finite set to itself is a surjection, so ϕ_a is a permutation of $\mathbb{Z}_n^*$. $\square$

DEFINITION. Order, $o_p(a)$ Let p be a prime number and $a \in \mathbb{Z}$ not a multiple of p. The order of a in $\mathbb{Z}_p$ is the least $k \in \mathbb{N}^+$ such that $a^k \equiv 1 \mod p$. We write the order of a in $\mathbb{Z}_p$ as $o_p(a)$.

If a is a multiple of p, then the order of a in $\mathbb{Z}_p$ is undefined since $a \equiv 0 \mod p$, and for all $k \in \mathbb{N}^+$,

$$a^k \equiv 0 \mod p.$$

The following proposition shows in particular that if a is not a multiple of p, then the order is well-defined (i.e., that there is some k with $a^k \equiv 1 \mod p$).

PROPOSITION 7.17. *Let $a \in \mathbb{Z}$, and p be a prime number such that $p \nmid a$. Then $o_p(a) < p$.*

PROOF. Let p be a prime number and $a \in \mathbb{Z}$ be such that a is not a multiple of p. By Lemma 7.5, as $p \nmid a$, then $p \nmid a^n$, and therefore $[a^n] \in \mathbb{Z}_p^*$ for any $n \in \mathbb{N}$. Since $| \mathbb{Z}_p^* | = p - 1$, the finite sequence

$$\langle [a^n] \mid 1 \le n \le p \rangle$$

must have a repetition. Let $1 \le n < k \le p$ be such that

$$a^n \equiv a^k \mod p.$$

Then

$$p \mid a^k - a^n.$$

Hence,

$$p \mid a^n(a^{k-n} - 1).$$

However, $p \nmid a^n$ and thus, by Proposition 7.3,

$$p \mid a^{k-n} - 1.$$

Thus,

$$a^{k-n} \equiv 1 \mod p.$$

Therefore,

$$o_p(a) \le k - n < p.$$

$\square$

PROPOSITION 7.18. *Let $a \in \mathbb{Z}$ and p be a prime number such that a is not a multiple of p. Then the remainder classes $[1], [a], [a^2], \ldots, [a^{o_p(a)-1}]$ in $\mathbb{Z}_p$ are distinct.*

PROOF. Exercise. $\square$

NOTATION. $S_a(n)$ Fix a prime p for the remainder of this section. Let a be an integer such that $p \nmid a$. Then for any positive natural number n, we let $S_a(n)$ denote the set of equivalence classes $\{[n \cdot a^k] \mid k \in \mathbb{N}\}$ in $\mathbb{Z}_p$. (Although $S_a(n)$ depends on the choice of p, we suppress this in the notation and assume that p is understood).

LEMMA 7.19. *Let $a \in \mathbb{Z}$ be such that $p \nmid a$. If $n \in \mathbb{N}^+$ is not a multiple of p, then*

$$|\, S_a(n)\, | \;=\; o_p(a).$$

PROOF. By Proposition 7.17, $o_p(a) < p$. Let $k = o_p(a)$. By Proposition 7.18 the remainder classes $[1], [a], [a^2], \ldots, [a^{k-1}]$ are distinct. Let ϕ_n be defined as in Lemma 7.16. Then ϕ_n is a permutation of $\mathbb{Z}_p^*$. Therefore the remainder classes $[n], [na^2], \ldots, [na^{k-1}]$ are distinct. But

$$na^k \equiv n \quad \mathrm{mod}\ p,$$

so

$$S_a(n) \;=\; \{[n], [na^2], \ldots, [na^{k-1}]\}$$

(why?). Therefore,

$$|\, S_a(n)\, | \;=\; o_p(a).$$

$\square$

LEMMA 7.20. *Let $a \in \mathbb{Z}$ be such that $p \nmid a$. Then for any $m, n \in \mathbb{N}^+$ that are not multiples of p, the sets $S_a(m)$ and $S_a(n)$ are either equal or disjoint.*

PROOF. Suppose $S_a(m) \cap S_a(n) \neq \emptyset$. Let $m, n \in \mathbb{N}$, $\gcd(m, p) = 1$, $\gcd(n, p) = 1$ and

$$[ma^i] \in S_a(n).$$

Then there is $j \in \mathbb{N}$ such that

$$[ma^i] = [na^j].$$

We may assume that $i < j$ since there are infinitely many $j \in \mathbb{N}^+$ that satisfy the equation. Then

$$[m] = [na^{j-i}].$$

So

$$[m] \in S_a(n).$$

Therefore, if $S_a(m)$ and $S_a(n)$ are not disjoint, we have

$$S_a(m) \subseteq S_a(n).$$

By symmetry, we also have

$$S_a(n) \subseteq S_a(m),$$

and so either

$$S_a(m) = S_a(n)$$

or

$$S_a(m) \cap S_a(n) = \emptyset.$$

$\square$

THEOREM 7.21. Fermat's little theorem *If $a \in \mathbb{Z}$ and p is a prime number such that $p \nmid a$, then*

$$a^{p-1} \equiv 1 \mod p.$$

PROOF. Let $k = o_p(a)$. We show that $k \mid (p-1)$. Let $n \in \mathbb{N}$, where n is not a multiple of p. By Lemma 7.19,

$$|S_a(n)| = k.$$

By Lemma 7.20, the sets

$$\{S_a(n) \mid n \in \mathbb{N}^+,\ p \nmid n\}$$

partition $\mathbb{Z}_p^*$ into sets of cardinality k. Therefore k divides $\mid Z_p^* \mid$. Since $\mid Z_p^* \mid = p-1$, we have

$$k \mid (p-1).$$

It follows that there is $j \in \mathbb{N}$ such that

$$a^{p-1} \equiv (a^k)^j \equiv 1^j \equiv 1 \mod p.$$

$\square$

COROLLARY 7.22. *If $a \in \mathbb{Z}$ and p is a prime number such that $p \nmid a$, then*

$$a^p \equiv a \mod p.$$

Fermat's little theorem is an important result in the theoretical study of prime numbers and for determining primality. How might the theorem be used? Consider the problem of deciding whether a particular natural number n is prime. To determine whether n is prime, you might invoke the Fundamental Theorem of Arithmetic and begin checking all the prime numbers up to $\sqrt{n}$ to determine whether any are nontrivial factors of n. We need not check primes greater than $\sqrt{n}$ since the existence of such a factor entails the existence of a factor less then $\sqrt{n}$, and by the Fundamental Theorem of Arithmetic, a prime factor less than $\sqrt{n}$. This may require checking many candidates—in addition to requiring that you *know* all of the prime numbers smaller than $\sqrt{n}$ or are willing to check factors that are not prime. For large n, this is a formidable challenge. Alternatively, you can seek $a \in \mathbb{Z}$ such that $[a^n] \neq [a]$ in $\mathbb{Z}_n$ in order to determine that n is not prime.

For instance, is 12,871 prime? We assume that you have access to a computer (doing these computations by hand can be tedious). One approach is to check for factors among the prime numbers less than $\sqrt{12,871}$, that is the 30 prime numbers less than 114. Alternatively, for $a \in \mathbb{Z}$, we can check whether

$$a^{12,871} \equiv a \mod 12,871.$$

If the answer is "no," then 12,871 is not prime. We shall try $a = 2$:

$$2^{12,871} \equiv 5,732 \mod 12,871.$$

Therefore 12,871 is not prime. If you were to check primes sequentially, you would have to check 18 primes before finding that 61 is the smallest prime that divides 12,871.

If $a^{12,871} \equiv a \mod 12,871$ for a given choice of a, then we can draw no conclusion. In fact there are nonprime numbers, n, such that for any choice of a,

$$a^n \equiv a \mod n.$$

Numbers that satisfy the conclusion of Theorem 7.21 but are not prime are called Carmichael numbers. So Fermat's little theorem can be used to show that a number is not prime but not to prove that a number is prime.

7.5 Divisibility and Polynomials

We apply some of the ideas on divisibility introduced in earlier sections of this chapter to polynomials with real coefficients, $\mathbb{R}[x]$. This requires us to treat polynomials algebraically. We begin by formally defining operations on $\mathbb{R}[x]$. Let $f, g \in \mathbb{R}[x]$,

$$f(x) = \sum_{n=0}^{N} a_n x^n$$

and

$$g(x) = \sum_{m=0}^{M} b_m x^m.$$

So f is a polynomial of degree at most N and g is a polynomial of degree at most M. To simplify our expressions, we subscribe to the convention that for the polynomials f and g, $a_n = 0$ for all $n > N$, and $b_m = 0$ for all $m > M$. That is, we may consider a polynomial as a power series in which all but finitely many of the coefficients equal 0.

REMARK. If a polynomial is identically equal to a nonzero constant, we say that the polynomial has degree zero. If the polynomial is identically zero, we do not define its degree. This is a notational convenience: a polynomial of degree zero is a nonzero constant.

We define addition and multiplication in $\mathbb{R}[x]$ by

$$f(x) + g(x) := \sum_{i=0}^{\max(M,N)} (a_i + b_i)x^i$$

and

$$f(x) \cdot g(x) := \sum_{i=0}^{M+N} \left(\sum_{j=0}^{i} a_j \cdot b_{i-j} \right) x^i.$$

You should confirm that $0 \in \mathbb{R}[x]$ is the additive identity in $\mathbb{R}[x]$ and that $1 \in \mathbb{R}[x]$ is the multiplicative identity in $\mathbb{R}[x]$. You should also verify that addition and multiplication in $\mathbb{R}[x]$ are

(1) associative
(2) commutative
(3) distributive (i.e., multiplication distributes over addition)

We shall prove that a version of the division algorithm holds for polynomials. Indeed, it is the reason that long division of polynomials is essentially similar to division of integers.

THEOREM 7.23.   *Division algorithm*   *If $f, g \in \mathbb{R}[x]$, and $g \neq 0$, then there are unique polynomials q and r such that*

$$f \; = \; q \cdot g + r$$

and either $r \; = \; 0$ or the degree of r is less than the degree of g.

DISCUSSION. *We argue first for the existence of a quotient and remainder satisfying the statement of the theorem. We let g be an arbitrary real polynomial and argue by induction on the degree of f — for this particular divisor g. The induction principle will yield the result for the divisor g and any dividend. Since g is an arbitrary real polynomial, the existence of a quotient and remainder is guaranteed for any divisor and dividend. Uniqueness is proved directly.*

PROOF. Let $g \in \mathbb{R}[x]$. If g is a constant, then $q(x) \; = \; (1/g(x))(f(x))$ and $r \; = \; 0$ satisfy the statement of the theorem. Furthermore, any remainder must be the zero polynomial since it is impossible to have degree smaller than the degree of g. Hence, $q(x) \; = \; (1/g(x))(f(x))$ is the unique quotient that satisfies the division algorithm.

Let g be a polynomial of degree greater than 0. We prove the result for all possible f (for this particular g) by induction on the degree of f. Let M be the degree of g and N be the degree of f.

BASE CASE: $N < M$

Then $q \; = \; 0$ and $r \; = \; f$ satisfy the conclusion of the theorem.

INDUCTION STEP: Let $N \geq M$ and assume that the result holds for all polynomials of degree less than N. We show that it holds for $f \in \mathbb{R}[x]$ of degree N. We assume that

$$f(x) \; = \; \sum_{n=0}^{N} a_n x^n$$

where $a_n \in \mathbb{R}$ (for $0 \le n \le N$) and $a_N \ne 0$. Let

$$g(x) = \sum_{m=0}^{M} b_m x^m$$

where $b_m \in \mathbb{R}$ (for $0 \le m \le M$) and $b_M \ne 0$. Let

$$h(x) = \left(\frac{a_N}{b_M}\right) x^{(N-M)}.$$

Then the degree of $f - h \cdot g$ is less than N or $f - h \cdot g$ is identically 0.
So there is $s \in \mathbb{R}[x]$ such that

$$f = h \cdot g + s$$

where $s = 0$ or the degree of s is less than N. If $s = 0$, then let
$q = h$ and $r = 0$.

Otherwise, by the induction hypothesis, there is some polynomial
$\bar{q}$ such that

$$s = \bar{q} \cdot g + r$$

where $r = 0$ or the degree of r is less than M. Thus

$$f = hg + s = hg + \bar{q}g + r = (h + \bar{q})g + r.$$

If we let $q = h + \bar{q}$ then

$$f = qg + r.$$

So, by the principle of induction, for any $f \in \mathbb{R}[x]$, there are q and r
such that

$$f = q \cdot g + r.$$

Since g was an arbitrary polynomial of degree greater than 0, the result
holds for all f and g.

We prove that q and r are unique. Let

$$f = qg + r = \bar{q}g + \bar{r}$$

where the remainders, r and $\bar{r}$, have degree less than the degree of g or are the zero polynomial. Then

$$qg + r - (\bar{q}g + \bar{r}) \;=\;$$
$$(q - \bar{q})g + (r - \bar{r}) \;=\; 0.$$

Let $Q = q - \bar{q}$ and $R = r - \bar{r}$. Assume that $Q \neq 0$. Then the degree of $Q \cdot g$ is no less than the degree of g. However the remainders r and $\bar{r}$ have degree less than the degree of g, or are the zero polynomial. Thus the degree of R is strictly less than the degree of g, or $R = 0$. The sum of two polynomials of different degree cannot be identically 0. Hence it is impossible that $Q \neq 0$. If $Q = 0$ then $R = 0$. Therefore

$$q = \bar{q}$$

and

$$r = \bar{r}$$

and the quotient and remainder are unique. □

COROLLARY 7.24. *Let $f \in \mathbb{R}[x]$ and $x_0 \in \mathbb{R}$. Then there is $q \in \mathbb{R}[x]$ such that*

$$f(x) \;=\; (x - x_0) \cdot q(x) + f(x_0).$$

PROOF. Apply the division algorithm with $g(x) = x - x_0$. Then the remainder r is of degree 0, or identically zero, so is constant, and evaluating

$$f(x) \;=\; (x - x_0)q(x) + r(x)$$

at $x = x_0$ gives $r(x) = f(x_0)$. Therefore,

$$f(x) \;=\; (x - x_0)q(x) + f(x_0).$$

□

We use these results to prove an algebraic property of polynomials.

DEFINITION.   Ideal If $I \subseteq \mathbb{R}[x]$ and $I \neq \emptyset$, then we call I an ideal of $\mathbb{R}[x]$ provided the following conditions are satisfied:

(1) If $f, g \in I$, then $f + g \in I$.
(2) If $f \in I$ and $g \in \mathbb{R}[x]$, then $f \cdot g \in I$.

An ideal of $\mathbb{R}[x]$ is a set that is closed under addition of elements in the ideal and multiplication by all elements of $\mathbb{R}[x]$, whether or not they are in the ideal. If you look closely at the definition of integer combination (Section 7.1), you will observe that the set of integer combinations of a pair of integers is closed under addition of elements in the set and multiplication by arbitrary integers. Of course this analogy between the integers and the polynomials is not accidental. If you generalize the idea of an integer combination to polynomials, you would say that the polynomial combinations of a pair of polynomials is an ideal of $\mathbb{R}[x]$. For the integers we were able to prove that the set of integer combinations of a pair of integers is precisely the integer multiples of the greatest common divisor of the integers. Can we prove an analogous result for polynomials?

DEFINITION.   Principal ideal An ideal I in $\mathbb{R}[x]$ is principal if there is $f \in \mathbb{R}[x]$ such that

$$I = \{f \cdot g \mid g \in \mathbb{R}[x]\}.$$

In the definition of principal ideal, f is called a generator of I. Theorem 7.7 can be restated to say that the set of integer combinations of a pair of integers is the principal ideal (in $\mathbb{Z}$) generated by the greatest common divisor of the pair.

THEOREM 7.25. *Every ideal of $\mathbb{R}[x]$ is principal.*

PROOF. Let I be an ideal of $\mathbb{R}[x]$. Let f be a polynomial of lowest degree in I. We prove that f generates I. Let $h \in I$. It is sufficient to show that h is a multiple of f. By Theorem 7.23, there are $q, r \in \mathbb{R}[x]$, $r = 0$ or the degree of r less than the degree of f, such that

$$h = qf + r.$$

Since I is an ideal and $f \in I$,

$$qf \in I$$

and

$$h - qf = r \in I.$$

By assumption f is of minimal degree in I, so $r = 0$. Therefore

$$h = qf$$

and f generates I. $\qquad\qquad\qquad\qquad\qquad\qquad\qquad\qquad$ □

This program seems to be moving us toward a result for polynomials that is analogous to the Fundamental Theorem of Arithmetic. A polynomial is irreducible if it cannot be written as the product of polynomials of lower degree. We shall prove in Theorem 9.44 that every polynomial in $\mathbb{R}[x]$ factors uniquely into the product of irreducible polynomials (up to the order of factors and multiplication by constants) and moreover that all irreducible polynomials are of degree at most 2.

Studying algebraic properties of polynomials is the most important motivating principle in algebra. Good texts on algebra include those by John Fraleigh [2] and Israel Herstein [3].

7.6 Exercises

EXERCISE 7.1. Let $n \in \mathbb{N}$. Prove that if n is not prime, then n has a prime factor $p \le \sqrt{n}$.

EXERCISE 7.2. Are $15,462,227$ and $15,462,229$ relatively prime?

EXERCISE 7.3. If $n \in \mathbb{N}$, under what conditions are n and $n + 2$ relatively prime?

EXERCISE 7.4. Prove that every natural number may be written as the product of a power of 2 and an odd number.

EXERCISE 7.5. Find $\gcd(8243235, 453169)$.

EXERCISE 7.6. Find $\gcd(15570555, 10872579)$.

EXERCISE 7.7. Let a and b be integers and $m = \gcd(a, b)$. Prove that $\frac{a}{m}$ and $\frac{b}{m}$ are relatively prime integers.

EXERCISE 7.8. Let a and b be positive integers with prime decomposition given by

$$a = \prod_{n=1}^{N} p_n^{r_n}$$

and

$$b = \prod_{n=1}^{N} p_n^{s_n}$$

where $p_n, r_n, s_n \in \mathbb{N}$ and p_n is prime for $1 \leq n \leq N$. Prove that if $t_n = \min(r_n, s_n)$ for $1 \leq n \leq N$, then

$$\gcd(a, b) = \prod_{n=1}^{N} p_n^{t_n}.$$

EXERCISE 7.9. In the statement of Lemma 7.16, suppose that $\gcd(a, n) \neq 1$. Prove that the function ϕ_a is not a permutation of $\mathbb{Z}_n^*$.

EXERCISE 7.10. Prove Proposition 7.18.

EXERCISE 7.11. Is 4,757 prime?

EXERCISE 7.12. Define an ideal of $\mathbb{Z}$ in the natural way: A set $I \subseteq \mathbb{Z}$ is an ideal of $\mathbb{Z}$ if for any $m, n \in I$ and $c \in \mathbb{Z}$,
1) $m + n \in I$
and
2) $mc \in I$.
If $a, b \in \mathbb{Z}$, prove that the set of integer combinations of a and b are an ideal of $\mathbb{Z}$.

EXERCISE 7.13. Prove that every ideal of $\mathbb{Z}$ is principal. (Hint: find the generator of the ideal.)

EXERCISE 7.14. Let p be prime and $\mathbb{Z}_p[x]$ be the set of polynomials with coefficients in $\mathbb{Z}_p$. What can you say about the roots of the polynomial $x^{p-1} - [1]$ in $\mathbb{Z}_p$? (We say that $[a] \in \mathbb{Z}_p$ is a root of a polynomial $f \in \mathbb{Z}_p[x]$ if $f([a]) = [0]$.)

EXERCISE 7.15. Prove that 0 is the additive identity in $\mathbb{R}[x]$ and 1 is the multiplicative identity in $\mathbb{R}[x]$. Use the formal definitions of addition and multiplication in $\mathbb{R}[x]$.

EXERCISE 7.16. Prove that the degree of the product of polynomials is equal to the sum of the degrees of the polynomials. Use the formal definition of multiplication in $\mathbb{R}[x]$.

EXERCISE 7.17. Let $p \in \mathbb{R}[x]$. Prove that p has an additive inverse in $\mathbb{R}[x]$. Prove that p has a multiplicative inverse iff p has degree 0. Use the formal definitions of addition and multiplication in $\mathbb{R}[x]$.

EXERCISE 7.18. Prove that addition and multiplication in $\mathbb{R}[x]$ are associative and commutative and that the distributive property holds. Use the formal definitions of addition and multiplication in $\mathbb{R}[x]$.

EXERCISE 7.19. For $0 \leq n \leq N$, let $a_n \in \mathbb{R}$. If $f = \sum_{n=0}^{N} a_n x^n$ and $g(x) = x - 1$, find the unique quotient and remainder where f is the dividend and g is the divisor.

EXERCISE 7.20. Let $f, g, q \in \mathbb{R}[x]$, $g \neq 0$. Suppose that f is the dividend, g the divisor, and q the quotient. Prove that the sum of the degree of g and the degree of q equals the degree of f.

EXERCISE 7.21. Is there a version of the Euclidean algorithm for $\mathbb{R}[x]$?

CHAPTER 8

The Real Numbers

What are the real numbers and why don't the rational numbers suffice for our mathematical needs? Ultimately the real numbers must satisfy certain axiomatic properties that we find desirable for interpreting the natural world while satisfying the mathematician's desire for a formal foundation for mathematical reasoning.

Of course the real numbers must contain the rational numbers. We also require that the real numbers satisfy rather obvious algebraic properties that hold for the rational numbers, such as commutativity of addition or the distributive property. These axioms allow us to use algebra to solve problems. Additionally we must satisfy geometric properties, like the triangle inequality, which permit the interpretation of positive real numbers as distances. We need our number system to contain numbers that arise from the algebraic and geometrical interpretation of numbers. Unfortunately the rational numbers do not suffice for this limited objective. For instance, $\sqrt{2}$, which you know by Example 3.23 to be irrational, arises geometrically as the length of the diagonal of the unit square and as the solution to the algebraic equation $x^2 = 2$.

The development of the limit gave rise to new questions about the real numbers. In particular, when are we assured that a sequence of numbers is convergent in our number system? The proof of convergence claims often use another property of the real numbers, the least upper bound property. Many of the powerful conclusions of calculus are consequences of this property. Loosely speaking, the least upper bound property implies that the real number line does not have any "holes." Put another way, if all the elements of one nonempty set of real numbers are less than all elements of another nonempty set of real numbers, then there is a real number greater than or equal to all the elements of the first set and less than or equal to all the elements of the second set. This property is called order-completeness and is formally defined in Section 8.10. Order-completeness, and its desirable consequences, do not hold for the rational numbers.

How do we prove the existence of a set with order and operations that satisfy all these needs simultaneously? One cannot simply assume that such a structure exists. It is possible that the properties specified are logically inconsistent. We might attempt to construct the set. What are the rules for the construction of a mathematical object? This question prompted mathematicians of the late nineteenth and early twentieth century to develop the rules for such a construction—the axioms of set theory.

For this reason we build the real numbers with a set-theoretic construction. That is, we shall construct the natural numbers, integers, rational numbers, and irrational numbers in turn, using basic sets, functions, and relations. In so doing we shall construct a set with order and operations that contains the rational numbers (or a structure that behaves precisely the way we expect the rational numbers to behave), satisfies the algebraic and order axioms, has the properties we need for calculus, and is constructed with the tools that you developed in Chapters 1 and 2.

8.1 The Natural Numbers

When we introduced the natural numbers in Chapter 1, we were explicit that we were not defining the set. Instead we proceeded under the assumption that you are familiar with the natural numbers by virtue of your previous mathematical experience. Now we define the natural numbers in the universe of sets, constructing them out of the empty set.

DEFINITION. Successor function Let Y be a set. The successor function, S, is defined by

$$S(Y) := Y \cup \{Y\}.$$

DEFINITION. Inductive set Let S be the successor function and X be any collection of sets satisfying the following conditions:

(1) $\emptyset \in X$.
(2) $[Y \in X] \Rightarrow [S(Y) \in X]$.

Then X is called an inductive set.

DEFINITION. Natural numbers Let X be any inductive set. The set of natural numbers is the intersection of all subsets of X that are inductive sets.

Are the natural numbers well-defined? That is, does the definition depend on the choice of the set X? If $\mathcal{F}$ is a family of sets, all of which are inductive, it is easily proved that the intersection over $\mathcal{F}$ is also inductive. If we are given sets X and Y that are inductive, will the sets give rise to the same set of "natural numbers"? Again it is easily seen that the answer is "yes" since $X \cap Y$ is a subset of both X and Y and is inductive. The "natural numbers" defined in terms of X and Y will be the "natural numbers" defined in terms of $X \cap Y$—they constitute the "smallest" inductive set. In order to define the natural numbers in the universe of sets, it must be granted that there exists an inductive set. It is an axiom of set theory that there is such a set, called the Axiom of Infinity (see Appendix B for a discussion of the axioms of set theory).

What does this set have to do with the natural numbers as we understand and use them in mathematics? Consider the function i defined by

$$i(0) = \emptyset$$

and

$$i(n+1) = i(n) \cup \{i(n)\}.$$

So

$$
\begin{aligned}
i(0) &= \emptyset \\
i(1) &= \{\emptyset\} \\
i(2) &= \{\emptyset, \{\emptyset\}\} \\
i(3) &= \{\emptyset, \{\emptyset\}, \{\emptyset, \{\emptyset\}\}\}.
\end{aligned}
$$

Then i gives a bijection between the natural numbers, as we understand them intuitively, and the minimal inductive set that we defined above.

Let us define $\ulcorner n \urcorner$ formally as the set one obtains by applying the successor function S to the empty set n times. So

$$0 = \emptyset$$

and for $n > 0$ the set

$$\ulcorner n \urcorner = \{\emptyset, \{\emptyset\}, \dots\}$$

has exactly n elements, and we shall identify it with the set

$$\{0, 1, \dots, n-1\}$$

that we earlier chose as the canonical set with n elements.

The set

$$\mathbf{N} := \{\ulcorner n \urcorner \mid n \in \mathbb{N}\} \tag{8.1}$$

is inductive, and therefore contains the natural numbers. Finally, the reader may confirm that $\mathbf{N}$ has no proper subset that is inductive.

To summarize the construction so far, the Axiom of Infinity guarantees that there is a set that is inductive. Pick such a set, X. The intersection of all subsets of X that are inductive is $\mathbf{N}$, which we can identify with the natural numbers (conceived intuitively) by the bijection i. In order to continue the construction, we consider $\mathbb{N}$ and $\mathbf{N}$ to be the same set. We need $\mathbb{N}$ to have the operations $+$ and $\cdot$ as well as the relation $\leq$.

We shall define addition in $\mathbb{N}$ with basic set operations and cardinality. If $m, n \in \mathbb{N}$, then we define addition by

$$m + n := \mid (\ulcorner m \urcorner \times \{\ulcorner 0 \urcorner\}) \cup (\ulcorner n \urcorner \times \{\ulcorner 1 \urcorner\}) \mid.$$

Recall that the cardinality of a finite set is the unique natural number that is bijective with the set—hence the complicated expression on the right hand side of the definition is a natural number. It is easy to confirm that addition defined in this manner agrees with the usual operation in $\mathbb{N}$. Why would we bother to define an operation you have understood for many years? We have defined addition of natural numbers as a *set operation*.

Multiplication is somewhat easier to define. If $m, n \in \mathbb{N}$, then

$$m \cdot n := \mid \ulcorner m \urcorner \times \ulcorner n \urcorner \mid.$$

(Of course, by $\ulcorner m \urcorner \times \ulcorner n \urcorner$ we mean the Cartesian product of the sets $\ulcorner m \urcorner$ and $\ulcorner n \urcorner$.) Finally if $m, n \in \mathbb{N}$

$$[m \leq n] \iff [\ulcorner m \urcorner \subseteq \ulcorner n \urcorner].$$

You should confirm that the operations and the relation agree with the usual $+$, $\cdot$, and $\leq$ on the natural numbers.

Having completed this construction, it is reasonable to ask whether $\mathbb{N}$ is truly the set of natural numbers. It is certainly justifiable for you to conclude that no clarity about the number 2 is provided by identifying it with the set $\{\emptyset, \{\emptyset\}\}$. What we gain is a reduction of numbers to sets that will carry us through the construction of all real numbers, including numbers that are not easy to construct.

8.2 The Integers

We construct the integers out of the natural numbers. The algebraic purpose of the integers is to include additive inverses for natural numbers. Of course this naturally gives rise to the operation of subtraction.

Let $Z = \mathbb{N} \times \mathbb{N}$. Define an equivalence relation, $\sim$ on Z by

$$\langle m_1, n_1 \rangle \sim \langle m_2, n_2 \rangle \quad \Longleftrightarrow \quad m_1 + n_2 = m_2 + n_1.$$

Then the integers are

$$\mathbf{Z} := Z/\sim.$$

We think of the ordered pair $\langle m, n \rangle \in \mathbf{Z}$ as being a representative of the integer $m - n$. We say that an integer is positive if $m > n$ and negative if $m < n$. It should be clear that the set of nonnegative integers (that is $\mathbb{N}$) is

$$\{[\langle m, n \rangle] \mid m \geq n\} = \{[\langle m, 0 \rangle] \mid m \in \mathbb{N}\}.$$

Let $\mathbb{Z}$ be the (intuitive) integers and let $i : \mathbf{Z} \to \mathbb{Z}$ be defined by

$$i([\langle m, n \rangle]) = m - n.$$

Then i is a bijection. As we did with the natural numbers, we shall construct operations and order on $\mathbf{Z}$ that agree with the usual operations and an order on $\mathbb{Z}$. Of course, we could use i and the usual definitions in $\mathbb{Z}$ to define operations and relations on $\mathbf{Z}$, but that would miss the spirit of the construction and would neglect the desire for set-theoretic definitions. Analogous to the construction of the previous section, we define $\mathbb{Z}$ as $\mathbf{Z}$. Let $x_1, x_2 \in \mathbb{Z}$ where $x_1 = [\langle m_1, n_1 \rangle]$ and $x_2 = [\langle m_2, n_2 \rangle]$. Addition is defined by

$$x_1 + x_2 = [\langle m_1 + m_2, n_1 + n_2 \rangle].$$

The additive inverse of $[\langle m, n \rangle]$ is $[\langle n, m \rangle]$ (i.e., the sum of these integers is $[\langle 0, 0 \rangle]$—the additive identity in $\mathbb{Z}$).

Multiplication is defined by

$$x_1 \cdot x_2 = [\langle m_1 \cdot m_2 + n_1 \cdot n_2, n_1 \cdot m_2 + m_1 \cdot n_2 \rangle].$$

The linear ordering on $\mathbb{Z}$ is defined by

$$x_1 \leq x_2 \quad \Longleftrightarrow \quad m_1 + n_2 \leq n_1 + m_2.$$

Addition and multiplication have been defined for the natural numbers, and the operations and linear ordering on $\mathbb{Z}$ are defined with respect to operations and the linear ordering that were previously defined for $\mathbb{N}$. Note that all our definitions were given in terms of representatives of equivalence classes. To show that $+$, $\cdot$, and $\leq$ are well-defined, we must show that the definitions are independent of the choice of representative—see Exercise 8.6.

8.3 The Rational Numbers

Rational numbers are ratios of integers, or nearly so. Of course, different numerators and denominators can give rise to the same rational number—indeed a good deal of elementary-school arithmetic is devoted to determining when two distinct expressions for rational numbers are equal. We built the integers from the natural numbers with equivalence classes of "differences" of natural numbers. We construct the rational numbers from the integers analogously, with equivalence classes of "quotients" of integers. Algebraically this gives rise to division.

Let $Q = \mathbb{Z} \times \mathbb{N}^+$. We define an equivalence relation $\sim$ on Q. If $\langle a, b \rangle, \langle c, d \rangle \in Q$, then

$$\langle a, b \rangle \sim \langle c, d \rangle \quad \Longleftrightarrow \quad a \cdot d = b \cdot c.$$

We define the rational numbers, $\mathbf{Q}$, as the equivalence classes of Q with respect to the equivalence relation $\sim$. That is,

$$\mathbf{Q} := Q/\sim .$$

We associate the equivalence classes of $\mathbf{Q}$ with the intuitive rational numbers via the bijection $i : \mathbb{Q} \to \mathbf{Q}$ defined by

$$i\left(\frac{p}{q}\right) \;=\; [\langle p, q \rangle]$$

for $\langle p, q \rangle \in Q$.

We define the operations and linear ordering on $\mathbb{Q}$ in terms of the operations and linear ordering in $\mathbb{Z}$. Define addition by

$$[\langle a, b \rangle] + [\langle c, d \rangle] := [\langle ad + bc, bd \rangle]$$

and multiplication by

$$[\langle a, b \rangle] \cdot [\langle c, d \rangle] := [\langle a \cdot c, b \cdot d \rangle].$$

We define the linear ordering on $\mathbb{Q}$ by

$$[\langle a, b \rangle] \leq [\langle c, d \rangle] \quad \text{iff} \quad a \cdot d \leq b \cdot c.$$

Through the construction of the rational numbers, we have used set operations to build mathematical structures with which you are already familiar. Consequently you are able to check that the construction behaves as you expect. For instance, one can easily prove that the operations we have constructed agree with the usual operations of addition and multiplication on the rational numbers. Similarly one can easily check that the relation we have constructed on $\mathbb{Q}$ agrees with the usual linear ordering of the rational numbers. Constructing the real numbers is more complicated.

8.4 The Real Numbers

We complete our construction of the real numbers (we have the irrational numbers remaining) with the objectives of proving the order-completeness of the real numbers and deriving some important consequences of completeness. Many of the most powerful and interesting results of calculus depend on this property of the real numbers. If you have been asked to accept some of these theorems on faith, now it is time to reward your trust.

There are a couple of different ways to construct the real numbers from the rational numbers. One approach is to define real numbers as convergent sequences of rational numbers. The other common approach is to characterize real numbers as subsets of rational numbers that satisfy certain conditions.

DEFINITION.   Dedekind cut   A Dedekind cut L is a nonempty proper subset of $\mathbb{Q}$ that has no maximal element and satisfies

$$(\forall\, a, b \in \mathbb{Q})\, [a \in L \,\wedge\, b < a] \;\Rightarrow\; [b \in L].$$

Let L be a Dedekind cut. Then there is some rational number $a \in L$, and therefore all rational numbers less then a are in L. Let $R = \mathbb{Q} \setminus L$. Since $L \neq \mathbb{Q}$, there is $c \in R$ and every rational number greater than c is in R. It is clear that $\{L, R\}$ is a partition of $\mathbb{Q}$ and that every element of L is less than every element of R. So Dedekind cuts "split" the rational numbers. We shall associate each Dedekind cut with a real number located at the split on the real number line.

REMARK. To help our mental picture of what is going on, we think of L as all rational numbers to the left of some fixed real number α, that is, as $(-\infty, \alpha) \cap \mathbb{Q}$, and R as the rational numbers to the right, $[\alpha, \infty) \cap \mathbb{Q}$. Of course we do not yet know what exactly we mean by "the real number α," but this is the idea to keep in mind. Note that R will have a least element iff α is rational.

To understand how Dedekind cuts relate to numbers, we construct an injection from the rational numbers to the Dedekind cuts. Let $\mathcal{D}$ be the set of Dedekind cuts. We define an injection $i \colon \mathbb{Q} \to \mathcal{D}$ by

$$i(a) \;=\; \{b \in \mathbb{Q} \mid b < a\}.$$

The function i is a well-defined injection that informs us of how $\mathbb{Q}$ fits into $\mathcal{D}$.

We shall define order and operations on $\mathcal{D}$ so that they agree with the usual linear ordering and operations on $\mathbb{Q}$ that are inherited in $i[\mathbb{Q}]$. That is, we shall define the linear order, addition, and multiplication on $\mathcal{D}$ so that for $a, b \in \mathbb{Q}$,

(1)

$$[a \leq b] \iff [i(a) \leq i(b)]$$

(2)

$$i(a + b) = i(a) + i(b)$$

(3)

$$i(a \cdot b) = i(a) \cdot i(b)$$

If we can do this, we can think of $\mathcal{D}$ as an extension of $\mathbb{Q}$. How do we do it?

For $L, K \in \mathcal{D}$, we define the relation $\leq$ in $\mathcal{D}$ by

$$[L \leq K] \iff [L \subseteq K].$$

You should confirm that $\leq$ is a linear ordering of $\mathcal{D}$ and that the relation $\leq$ on $i[\mathbb{Q}]$ satisfies (1). If $L \in \mathcal{D}$ and $L < i(0)$, we say that L is negative. If $L > i(0)$, we say that L is positive.

With a similar objective in mind we define addition and multiplication on $\mathcal{D}$. That is, we want the operations to satisfy certain properties of addition and multiplication and we want the operations defined on $i[\mathbb{Q}]$ to agree with the operations on $\mathbb{Q}$.

If $L, K \in \mathcal{D}$, then

$$L + K := \{a + b \mid a \in L \text{ and } b \in K\}.$$

Verify that $L + K$ is a Dedekind cut, and that (2) holds.

Multiplication takes a bit more effort to define. (Why can't we let $L \cdot K = \{ab \,|\, a \in L, \, b \in K\}$?) If L or K is $i(0)$, then

$$L \cdot K \,:=\, i(0).$$

If $L, K \in \mathcal{D}$ are both positive, then

$$L \cdot K \,=\, \{a \cdot b \,|\, a \in L, \, b \in K, \, a > 0, \text{ and } b > 0\} \cup \{c \in \mathbb{Q} \,|\, c \leq 0\}.$$

Verify that $L \cdot K$ is a Dedekind cut, and that (3) holds for $a, b > 0$.

How do we define multiplication by "negative" Dedekind cuts? Let us start with defining multiplication by -1. Let $L \in \mathcal{D}$ and $R = \mathbb{Q} \backslash L$. We define $-L$ by

$$-L \,:=\, \{c \in \mathbb{Q} \,|\, (\exists r \in R) \ -c > r\}.$$

Now we can define multiplication on arbitrary elements of $\mathcal{D}$ to satisfy the properties we desire. If $L, K \in \mathcal{D}$ and both are negative, then

$$L \cdot K \,:=\, (-L \cdot -K).$$

If exactly one of L and K is negative, then

$$L \cdot K \,:=\, -(-L \cdot K).$$

DEFINITION.   Real numbers, $\mathbb{R}$ The real numbers are the Dedekind cuts, with addition, multiplication, and $\leq$ defined as above. We denote the real numbers by $\mathbb{R}$ when we do not need to think of them explicitly as Dedekind cuts.

We have defined the real numbers as sets of rational numbers. Since the rational numbers were defined using basic ideas about sets, functions and relations, so are the real numbers. The properties of the real numbers that we discussed at the beginning of this section are satisfied by the Dedekind cuts. For every rational number a, we identify a with the Dedekind cut $i(a)$.

THEOREM 8.2. *The real numbers as defined above satisfy the following:*

    (i) *Addition and multiplication are both commutative and associative.*

    (ii) $(\forall L \in \mathcal{D})\ L + 0 = L,\ L \cdot 1 = L.$

    (iii) $(\forall L \in \mathcal{D})\ L + (-L) = 0.$

    (iv) $(\forall L \in \mathcal{D} \setminus \{0\})(\exists K \in \mathcal{D})\ L \cdot K = 1.$

    (v) $(\forall L, K, J \in \mathcal{D})\ L \cdot (K + J) = L \cdot K + L \cdot J.$

PROOF. Exercise.   □

8.5 The Least Upper Bound Property

DEFINITION. Upper bound Let $X \subset \mathcal{D}$. We say that X is bounded above if there is $M \in \mathcal{D}$ such that

$$(\forall x \in X)\ x \le M.$$

In this event we say that M is an upper bound for X.

DEFINITION. Least upper bound Let $X \subset \mathcal{D}$ be bounded above. Suppose M is an upper bound for X such that for any upper bound N for X, $M \le N$. Then the number M is called the least upper bound for X.

Lower bound and greatest lower bound are defined analogously.

THEOREM 8.3. *Least upper bound property* *If X is a nonempty subset of $\mathcal{D}$ and is bounded above, then X has a least upper bound. If it is bounded below, then it has a greatest lower bound.*

PROOF. Let $X \subset \mathcal{D}$ be bounded above. Let

$$M = \bigcup_{L \in X} L \subseteq \mathbb{Q}.$$

The set M is bounded above (why?), and hence $M \ne \mathbb{Q}$. Any element of M is an element of some $L \in X$ and consequently cannot be a maximal element of L. Therefore M has no largest element. If $a \in M$, $c \in \mathbb{Q}$, and $c < a$, then $c \in M$. Therefore M is a Dedekind cut. For any $L \in X$, $L \subseteq M$ and hence

$$L \leq M.$$

That is, M is an upper bound for X.

Let $K < M$. Then there is $a \in M \setminus K$. So a is in some L_0 in X. Therefore L_0 is not contained in K and K is not an upper bound for X. It follows that M is the least upper bound for X.

We leave the argument for the existence of a greatest lower bound to the reader.   □

The least upper bound property is the essential property of real numbers that permits the main theorems of calculus. It is the reason we use this large set, rather than, say, the algebraic numbers. It uniquely characterizes the real numbers as an extension of the rational numbers—see Theorem 8.23 for a precise statement.

Now that we have proved this key property, we shall use $\mathbb{R}$ to denote the set of real numbers, identifying a real number α with the Dedekind cut $(-\infty, \alpha) \cap \mathbb{Q}$. We shall no longer need to concern ourselves with Dedekind cuts per se.

8.6 Real Sequences

Recall that a sequence is a function with domain $\mathbb{N}$ (or $\mathbb{N}^+$). A real sequence is a real-valued sequence (that is, the range of the sequence is a subset of the real numbers).

DEFINITION. Subsequence Let $\langle a_n \mid n \in \mathbb{N} \rangle$ be a sequence and $f \in \mathbb{N}^{\mathbb{N}}$ be a strictly increasing sequence of natural numbers. Then

$$\langle a_{f(n)} \mid n \in \mathbb{N} \rangle$$

is a subsequence of $\langle a_n \mid n \in \mathbb{N} \rangle$.

EXAMPLE 8.4. Let s be the sequence

$$\langle 2n \mid n \in \mathbb{N} \rangle \;=\; \langle 0, 2, 4, 6, 8, \ldots \rangle.$$

Then the sequence t given by

$$\langle 6n \mid n \in \mathbb{N} \rangle \;=\; \langle 0, 6, 12, 18, \ldots \rangle$$

is a subsequence of s. In this example, $f(n) \;=\; 3n$ is the function that demonstrates that t is a subsequence of s. Another subsequence of s is the sequence

$$\langle 2^{5n+3} \mid n \in \mathbb{N} \rangle.$$

Recall that a sequence $\langle a_n \rangle$ is called *nondecreasing* if $a_{n+1} \geq a_n$ for all n. It is called nonincreasing if the inequality is reversed. Everything that is true for a nondecreasing sequence is true, with inequalities reversed, for nonincreasing sequences (why?), so rather than state everything twice, we can use the word *monotonic* to mean a sequence that is either nonincreasing (everywhere) or nondecreasing.

LEMMA 8.5. *Every nondecreasing real sequence $\langle a_n \mid n \in \mathbb{N} \rangle$ that is bounded above converges to its least upper bound. Every nonincreasing real sequence that is bounded below converges to its greatest lower bound.*

PROOF. We shall only prove the first assertion. Let M be the least upper bound of $\langle a_n \rangle$. Let $\varepsilon > 0$. Since M is the least upper bound, there is $N \in \mathbb{N}$ such that,

$$0 < M - a_N < \varepsilon.$$

Since the sequence is nondecreasing,

$$(\forall n \geq N)\; 0 < M - a_n < \varepsilon.$$

Therefore M is the limit of the sequence, as desired. $\square$

THEOREM 8.6. *Bolzano-Weierstrass theorem* *Let $[b, c]$ be a closed bounded interval of real numbers and $s = \langle a_n \mid n \in \mathbb{N} \rangle$ be a sequence of real numbers such that*

$$(\forall n \in \mathbb{N}) \ a_n \in [b, c].$$

Then $\langle a_n \mid n \in \mathbb{N} \rangle$ has a convergent subsequence with limit in $[b, c]$.

DISCUSSION. We consider a nested sequence of intervals, all of which contain infinitely many elements of the range of the sequence s, with the radius of the intervals approaching 0. We construct a subsequence of s by sequentially selecting elements in the intersection of the range of s and the successive intervals. We then show that the subsequence we construct is convergent.

PROOF. We prove the theorem for the closed unit interval $[0, 1]$. It is straightforward to generalize this argument to arbitrary closed bounded intervals.

If the range of the sequence is a finite set, then at least one element of the range, a_n, must have an infinite pre-image. The pre-image of a_n gives a subsequence that converges to a_n. Therefore we assume that the range of the sequence is infinite. Let S be the range of the sequence $\langle a_n \rangle$.

We define a nested sequence of closed intervals, $I_n = \langle [b_n, c_n] \mid n \in \mathbb{N} \rangle$ satisfying

(1) $I_0 = [0, 1]$
(2) for all $n \in \mathbb{N}$, $I_{n+1} \subset I_n$
(3) $c_n - b_n = \frac{1}{2^n}$
(4) for all $n \in \mathbb{N}$, $I_n \cap S$ is infinite

Let $I_0 = [0, 1]$. Assume that we have I_n satisfying the conditions above. At least one of the intervals $[b_n, b_n + \frac{1}{2^{n+1}}]$ and $[b_n + \frac{1}{2^{n+1}}, c_n]$ must contain infinitely many elements of S. Let $I_{n+1} = [b_n, b_n + \frac{1}{2^{n+1}}]$

if the intersection of this set with S is infinite; otherwise let $I_{n+1} =$ $[b_n + \frac{1}{2^{n+1}}, c_n]$. Then I_{n+1} satisfies the conditions above.

The sequence of left endpoints of the intervals I_n, $\langle b_n \mid n \in \mathbb{N} \rangle$ is nondecreasing. The sequence of right endpoints of the intervals I_n, $\langle c_n \mid n \in \mathbb{N} \rangle$ is nonincreasing. Furthermore, for any $m, n \in \mathbb{N}$,

$$b_m < c_n.$$

The set $\{b_n \mid n \in \mathbb{N}\}$ is bounded above, so by the least upper bound property the set has a least upper bound, β. Similarly the set $\{c_n \mid n \in \mathbb{N}\}$ has a greatest lower bound γ. By Lemma 8.5

$$\lim_{n \to \infty} b_n = \beta$$
$$\lim_{n \to \infty} c_n = \gamma.$$

By the triangle inequality, for any $n \in \mathbb{N}$,

$$|\beta - \gamma| \leq |\beta - b_n| + |b_n - c_n| + |c_n - \gamma|.$$

All three terms on the right-hand side of the inequality tend to 0 as n approaches infinity, so for any $\varepsilon > 0$,

$$|\beta - \gamma| < \varepsilon.$$

Hence $\beta = \gamma$.

We now want to define a subsequence that converges to β by choosing a point in each interval I_n in turn. Formally we do this by defining $f \in \mathbb{N}^\mathbb{N}$ recursively by

$$f(0) = 0$$

and $f(n+1)$ is the least $k \in \mathbb{N}$ such that

$$[k > f(n)] \wedge [a_k \in I_{n+1}].$$

This is well-defined since $S \cap I_{n+1}$ is infinite. Then the sequence $\langle a_{f(n)} \mid n \in \mathbb{N} \rangle$ converges to β. To see this, let $\varepsilon > 0$. For any $n \in \mathbb{N}$ such that $\frac{1}{2^n} < \varepsilon$,

$$|\beta - a_{f(n)}| < c_n - b_n = \frac{1}{2^n} < \varepsilon.$$

Therefore $\langle a_{f(n)} \mid n \in \mathbb{N} \rangle$ is a convergent subsequence converging to β. $\qquad\qquad\qquad\qquad\qquad\qquad\qquad\qquad\qquad\qquad\qquad\square$

DEFINITION. ~Cauchy sequence~ Let $\langle a_n \mid n \in \mathbb{N} \rangle$ be a sequence. The sequence $\langle a_n \rangle$ is a Cauchy sequence if

$$(\forall \varepsilon > 0)(\exists N \in \mathbb{N})(\forall m, n \in \mathbb{N}) \, [m, n \geq N] \Rightarrow [\mid a_m - a_n \mid < \varepsilon].$$

THEOREM 8.7. *A real sequence converges iff it is a Cauchy sequence.*

PROOF. $\Rightarrow$

Let $\langle a_n \mid n \in \mathbb{N} \rangle$ be a sequence of real numbers that converges to $a \in \mathbb{R}$. Let $\varepsilon > 0$ and $N \in \mathbb{N}$ be such that

$$(\forall n \geq N) \mid a - a_n \mid < \frac{\varepsilon}{2}.$$

Then for any $m, n \geq N$,

$$\mid a_n - a_m \mid \leq \mid a_n - a \mid + \mid a - a_m \mid < \frac{\varepsilon}{2} + \frac{\varepsilon}{2} = \varepsilon.$$

Therefore $\langle a_n \mid n \in \mathbb{N} \rangle$ is a Cauchy sequence.

$\Leftarrow$

Let $\langle a_n \mid n \in \mathbb{N} \rangle$ be a Cauchy sequence. Then

$$(\exists N \in \mathbb{N})(\forall m, n > N) \mid a_n - a_m \mid < 1.$$

Every term in the sequence after the Nth term is in the ε-neighborhood of a_N. So

$$(\forall n \geq N) \, a_n \in [a_N - 1, a_N + 1].$$

The sequence $\langle a_n \mid n \geq N \rangle$ satisfies the hypotheses of the Bolzano-Weierstrass theorem and thus has a convergent subsequence.

Let $\langle a_{f(n)} \mid n \in \mathbb{N} \rangle$ be a convergent subsequence of $\langle a_n \mid n \in \mathbb{N} \rangle$ converging to $a \in \mathbb{R}$. Let $\varepsilon > 0$. Since $\langle a_n \rangle$ is Cauchy, there is N_1 such that

$$(\forall m, n \geq N_1) \mid a_m - a_n \mid < \frac{\varepsilon}{2}.$$

Furthermore, there is $N_2 \in \mathbb{N}$ such that

$$(\forall n \geq N_2) \ |\, a_{f(n)} - a \,| < \frac{\varepsilon}{2}.$$

Let $N_3 \geq N_1, f(N_2)$. Then $N_3 \geq N_2$ and

$$(\forall n \geq N_3) \ |\, a_n - a \,| \leq |\, a_n - a_{f(n)} \,| + |\, a_{f(n)} - a \,| < \varepsilon.$$

Therefore the sequence $\langle a_n \mid n \in \mathbb{N} \rangle$ converges to a. $\square$

Cauchy sequences get at the essence of the order-completeness of the real numbers. A Cauchy sequence of rational numbers need not converge to a rational number. For instance, let a be any irrational number, and let a_n be the decimal approximation of a to the nth digit. The sequence $\langle a_n \rangle$ is a Cauchy sequence of rational numbers that converges to an irrational number. However if a Cauchy sequence fails to converge in a set of numbers, it is reasonable to say that there is a gap in the set of numbers. The real numbers are defined so that these gaps are filled.

8.7 Ratio Test

One of the uses of the order-completeness of the real numbers is proving that an infinite sequence converges, without having to know much about the number to which it converges. In Chapter 5 we allude to the ratio test in claiming that the Taylor polynomial for the exponential function evaluated at a real number a, $\sum_{k=0}^{\infty} \frac{a^k}{k!}$ converges. How do we prove that an infinite sum converges? If we have an idea of its limit, we might show that the sequence of partial sums approaches this value. This is how we prove that the geometric sum with ratio less than one converges. Many important mathematical functions are defined by infinite sums, and the limit of the sum defines the value of the function. In this case we need to show that the sum converges using properties of the real numbers.

DEFINITION. Absolute convergence Let $\langle a_n \rangle$ be an infinite sequence. If the infinite sum

$$\sum_{k=0}^{\infty} |a_k|$$

converges, then the infinite sum $\sum_{k=0}^{\infty} a_k$ is said to converge absolutely.

LEMMA 8.8. *If an infinite sum converges absolutely, then it converges.*

PROOF. Assume $\sum_{k=0}^{\infty} a_k$ converges absolutely. We show that the sequence of partial sums of this series, $\langle s_n \mid n \in \mathbb{N} \rangle$, is a Cauchy sequence. For $n \in \mathbb{N}$, let

$$b_n = |a_n|.$$

Then $\sum_{k=0}^{\infty} b_k$ converges. Let $\langle t_n \mid n \in \mathbb{N} \rangle$ be the sequence of partial sums of $\sum_{k=0}^{\infty} b_k$. By Theorem 8.7, $\langle t_n \rangle$ is a Cauchy sequence. Let $\varepsilon > 0$. Then there is $N \in \mathbb{N}$ such that for any $n \geq m \geq N$,

$$|t_n - t_m| \leq \varepsilon.$$

By a generalization of the triangle inequality (see Exercise 8.24),

$$|s_n - s_m| = \left| \sum_{k=m+1}^{n} a_k \right| \leq \sum_{k=m+1}^{n} b_k = |t_n - t_m| < \varepsilon.$$

Hence $\langle s_n \rangle$ is a Cauchy sequence and converges. Therefore $\sum_{k=0}^{\infty} a_k$ converges. □

THEOREM 8.9. *Ratio test* *Suppose $\langle a_k \rangle$ is an infinite sequence of real numbers and that there is $N \in \mathbb{N}$ and a positive real number $r < 1$ such that for all $n \geq N$,*

$$\left| \frac{a_{n+1}}{a_n} \right| \leq r.$$

Then $\sum_{k=0}^{\infty} a_k$ converges.

PROOF. Let $\sum_{k=0}^{\infty} a_k$ be an infinite sum with terms satisfying the hypothesis. For $n \in \mathbb{N}$, let $b_n = |a_n|$. By assumption, there is $N \in \mathbb{N}$ and a positive real number $r < 1$ such that for all $n \geq N$,

$$\frac{b_{n+1}}{b_n} \leq r.$$

We may assume without loss of generality that $N = 0$ since the series $\sum_{k=0}^{\infty} b_k$ converges iff $\sum_{k=N}^{\infty} b_k$ converges, and if necessary we may ignore finitely many terms of the infinite sum. We claim that for all $n \in \mathbb{N}$,

$$b_n \leq b_0 r^n.$$

If $n = 0$, the claim is obvious. Assume the claim holds at n. By assumption,

$$\frac{b_{n+1}}{b_n} \leq r.$$

Therefore,

$$b_{n+1} \leq r b_n \leq r b_0 r^n \leq b_0 r^{n+1}.$$

By Exercise 5.28, the geometric sum with radius $-1 < r < 1$ converges to $\frac{1}{1-r}$. Therefore, for any $n \in \mathbb{N}$,

$$s_n := \sum_{k=0}^{n} b_k \leq \sum_{k=0}^{n} b_0 r^k = b_0 \left(\sum_{k=0}^{\infty} r^k \right) \leq \frac{b_0}{1-r}.$$

The sequence of partial sums, $\langle s_n \rangle$, is a monotonic bounded sequence and by Lemma 8.5, converges. Therefore $\sum_{k=0}^{\infty} a_k$ converges absolutely. By Lemma 8.8 the sum converges. □

8.8 Real Functions

If you reread your calculus text, you will observe that many of the theorems of calculus are ultimately dependent on the Intermediate Value theorem.

THEOREM 8.10.   *Intermediate Value theorem*   *Let f be a continuous real function on a closed bounded interval $[a, b]$. If $f(a) < L < f(b)$ or $f(b) < L < f(a)$, then*

$$(\exists\ c \in (a, b))\quad f(c) \;=\; L.$$

PROOF. Let f be a continuous real function on a closed bounded interval $[a, b]$, and let $f(a) < L < f(b)$. We prove the special case $L \;=\; 0$. Given the result for $L \;=\; 0$, the theorem follows from application of the special case to the function $f(x) - L$.

Let

$$X \;=\; \{x \in [a, b] \mid (\forall\, y \in [a, x])\ f(y) \le 0\}.$$

Then $X \ne \emptyset$ and X is bounded above by b. By the least upper bound property, X has a least upper bound, $m \le b$. The function f is continuous, and hence $\lim_{x \to m} f(x) \;=\; f(m)$. If $f(m) \;=\; 0$, the theorem is proved.

(i) Assume that $f(m) > 0$. Let $0 < \varepsilon < f(m)$. For any $x \in [a, m)$, $f(x) \le 0$ and

$$|\, f(x) - f(m)\,| \ge f(m) > \varepsilon.$$

Consequently for any $\delta > 0$, there is x in the punctured δ-neighborhood of m such that

$$|\, f(x) - f(m)\,| \ge \varepsilon.$$

This contradicts the assumption that $\lim_{x \to m} f(x) \;=\; f(m)$. Therefore $f(m) \le 0$.

(ii) Assume that $f(m) < 0$. Let $0 < \varepsilon < |\, f(m)\,|$. For any $\delta > 0$, there is $x \in (m, m + \delta)$ such that $f(x) > 0$. Otherwise

$$[a, m + \delta) \subseteq X,$$

contradicting the assumption that m is the least upper bound for X. So for any $\delta > 0$, there is x in the punctured δ-neighborhood of m such that

$$| f(x) - f(m) | \geq | f(m) | > \varepsilon.$$

This contradicts the assumption that f is continuous at m. Therefore $f(m) = 0$.

$\square$

THEOREM 8.11. *Extreme Value theorem* *If f is a continuous real function on a closed bounded interval $[a, b]$, then f achieves a maximum and a minimum on $[a, b]$.*

PROOF. We show first that the range of $f|_{[a,b]}$ is bounded above and below. By way of contradiction, suppose that the range of f is not bounded above. For $n \in \mathbb{N}$, let $a_n \in [a, b]$ be such that $f(a_n) > n$. By the Bolzano-Weierstrass theorem, the sequence $\langle a_n \rangle$ has a convergent subsequence, $\langle a_{g(n)} \rangle$, converging to some number $c \in [a, b]$. By the continuity of f, if $c \in (a, b)$ then

$$f(c) = \lim_{x \to c} f(x) = \lim_{n \to \infty} f(a_{g(n)}).$$

(See Exercise 8.25.) If c is an endpoint of $[a, b]$, we make the corresponding claim for the appropriate one-sided limit. However, for any $n \in \mathbb{N}$,

$$f(a_{g(n)}) > g(n) > n.$$

Hence, $\lim_{n \to \infty} f(a_{g(n)})$ does not exist. Therefore the range of f is bounded above. Similarly, the range of f is bounded below. By the least upper bound property, the range of f has a least upper bound, M, and a greatest lower bound, L.

Since M is a least upper bound for the range of f, for any $\varepsilon > 0$, there is $x \in [a, b]$ such that

$$| f(x) - M | < \varepsilon.$$

For $n \in \mathbb{N}^+$, let $a_n \in [a, b]$ be such that

$$| f(a_n) - M | < \frac{1}{n}.$$

The sequence $\langle a_n \rangle$ has a convergent subsequence by the Bolzano-Weierstrass theorem. Let $\langle c_n \rangle$ be a convergent subsequence of $\langle a_n \rangle$ with limit $c \in [a, b]$. Since $\langle c_n \rangle$ is a subsequence of $\langle a_n \rangle$, for any $n \in \mathbb{N}^+$,

$$| f(c_n) - M | < \frac{1}{n}.$$

Hence,

$$\lim_{n \to \infty} f(c_n) = M.$$

By the continuity of f, if $c \in (a, b)$ then

$$\lim_{x \to c} f(x) = f(c) = \lim_{n \to \infty} f(c_n) = M.$$

If c is an endpoint of $[a, b]$, we have the analogous claim for the appropriate one-sided limit. Therefore f achieves a maximum value on $[a, b]$. By an analogous argument, f achieves a minimum value on $[a, b]$. □

By the Extreme Value theorem, a continuous function achieves extreme values on a closed bounded interval. It is easy to construct examples for which the theorem fails for open intervals. The Extreme Value theorem has in common with the least upper bound property that it guarantees the existence of a number satisfying a desirable condition without providing additional information about the number itself. Quite often it is enough to know abstractly that a function attains its extremum without having to further distinguish the object. What more can we conclude about the extreme values of a function?

THEOREM 8.12. *Let f be a real function defined on an interval (a, b). If $c \in (a, b)$ is such that $f(c)$ is an extreme value of f on (a, b) and f is differentiable at c, then $f'(c) = 0$.*

PROOF. Let f and c satisfy the hypotheses of the theorem. Suppose that $f(c)$ is the maximum value achieved by f on (a, b). For any $x \in (a, c)$, $f(x) \le f(c)$ and

$$\frac{f(c) - f(x)}{c - x} \ge 0.$$

Therefore,

$$\lim_{x \to c^-} \frac{f(c) - f(x)}{c - x} \ge 0.$$

Similarly,

$$\lim_{x \to c^+} \frac{f(c) - f(x)}{c - x} \le 0.$$

However f is differentiable at c, so

$$0 \le \lim_{x \to c^-} \frac{f(c) - f(x)}{c - x} = f'(c) = \lim_{x \to c^+} \frac{f(c) - f(x)}{c - x} \le 0.$$

A similar argument proves the claim for $f(c)$ a minimum value of f on (a, b). $\square$

COROLLARY 8.13. *Let f be a continuous real function on a closed bounded interval $[a, b]$. Then f achieves a maximum and minimum on $[a, b]$, and if $c \in [a, b]$ is a number at which f achieves an extreme value, then one of the following must be true of c:*

(i) $f'(c) = 0$.
(ii) *f is not differentiable at c.*
(iii) *c is an endpoint of $[a, b]$.*

THEOREM 8.14. *Mean Value theorem Let f be a continuous real function on a closed bounded interval $[a, b]$ and differentiable on (a, b). Then there is $c \in (a, b)$ such that*

$$f'(c) = \frac{f(b) - f(a)}{b - a}.$$

PROOF. We first prove a special case of the Mean Value theorem, known as Rolle's theorem. Assume that $f(a) = f(b)$. We prove that there is $x \in (a, b)$ such that $f'(x) = 0$.

If f is constant, then $f'(x) = 0$ for all $x \in (a, b)$. Assume that f is nonconstant and that there is $x \in (a, b)$ such that $f(x) > f(a)$. By the Extreme Value theorem, f achieves a maximum value M on $[a, b]$. Thus,

$$M > f(a) = f(b).$$

Let $c \in (a, b)$ be such that $f(c) = M$. By Theorem 8.12, $f'(c) = 0$. If there is $x \in (a, b)$ such that $f(x) < f(a)$, the proof is similar.

To prove the Mean Value theorem in general, we reduce it to Rolle's theorem. We subtract from $f(x)$ the line segment formed by $(a, f(a))$ and $(b, f(b))$. Let

$$g(x) = f(x) - f(a) - \frac{f(b) - f(a)}{b - a}(x - a).$$

The function $g(x)$ satisfies the hypotheses of Rolle's theorem. So there is $c \in (a, b)$ such that $g'(c) = 0$. Since

$$g'(x) = f'(x) - \frac{f(b) - f(a)}{b - a},$$

we have

$$g'(c) = f'(c) - \frac{f(b) - f(a)}{b - a} = 0$$

and

$$f'(c) = \frac{f(b) - f(a)}{b - a}.$$

□

The Mean Value theorem has many practical consequences, one of which we state here.

COROLLARY 8.15. *Let f be a differentiable function on (a, b). If $f'(x) > 0$ (respectively $f'(x) < 0$) on (a, b), then f is increasing (respectively decreasing) on (a, b).*

8.9 Cardinality of the Real Numbers

We finished Chapter 6 with the unproved claim that the real numbers
are uncountable. Now that we have a formal definition of the real numbers, we are ready to complete our investigation of the cardinality of $\mathbb{R}$.
By Theorem 6.9 the set of infinite decimal sequences is uncountable,
with cardinality $2^{\aleph_0}$. We went on to claim that this had consequences
for the cardinality of $\mathbb{R}$. We now consider the related question of the
cardinality of the closed unit interval $[0, 1]$.

PROPOSITION 8.16. $\mid [0, 1] \mid = \mid \mathbb{R} \mid$.

PROOF. Define $f : [0, \infty) \to (1/2, 1]$ by
$$ f(x) = \frac{1}{x + 2} + \frac{1}{2}. $$
Then f is an injection. Let $\mathbb{R}^-$ be the negative real numbers, and define
$g : \mathbb{R}^- \to [0, 1/2)$ by
$$ g(x) = \frac{-1}{x - 2}. $$
Then g is an injection. Let $h : \mathbb{R} \to [0, 1]$ be the union of the functions
f and g. Then h is clearly an injection. The identity function on $[0, 1]$
is an injection into $\mathbb{R}$. By the Schröder-Bernstein theorem,
$$ \mid [0, 1] \mid = \mid \mathbb{R} \mid. $$

□

We investigate the relationship between infinite decimal expansions
(which are related to infinite decimal sequences) and the real numbers.
We restrict our attention to infinite decimal expansions of numbers in
the unit interval $[0, 1]$.

DEFINITION. Infinite decimal expansion For all $n \in \mathbb{N}^+$, let $a_n \in$
$\ulcorner 10 \urcorner$. Then
$$ .a_1 a_2 \ldots a_n \ldots $$
is an infinite decimal expansion.

Let s be an infinite decimal expansion $.a_1 a_2 \ldots$. For $n \in \mathbb{N}$, let

$$s_n := .a_1 \ldots a_n = \sum_{k=1}^{n} a_k 10^{-k}.$$

We want to associate infinite decimal expansions with real numbers (understood as Dedekind cuts). We interpret infinite decimal expansions as Cauchy sequences of rational numbers.

Let D be the set of infinite decimal expansions, and let $f : D \to \mathbb{R}$ be defined by

$$f(.a_1 \ldots) = \lim_{n \to \infty} s_n.$$

The sequence $\langle s_n \rangle$ is a Cauchy sequence, so it converges to a real number. Let

$$L := \{ x \in \mathbb{Q} \mid (\exists n \in \mathbb{N}) \, x < s_n \}.$$

The set L is a Dedekind cut, and $f(s) = L$. That is

$$\lim_{n \to \infty} s_n = L.$$

L is the least upper bound of the set $\{ s_n \mid n \in \mathbb{N} \}$. We can associate with every infinite decimal expansion a real number in the unit interval and can thereby define a function $f : D \to [0, 1]$. Is f a surjection? That is, can every real number in the unit interval be realized as an infinite decimal expansion? Let $x \in [0, 1]$. We define an increasing sequence of rational numbers converging to x. For $n \in N^+$, let s_n be the largest decimal expansion to n decimal places that is no greater than x. If $n < m$, then s_n is a truncation of s_m. Let

$$s = \lim_{n \to \infty} s_n.$$

Then $f(s) = x$. Therefore f is a surjection onto $[0, 1]$.

It would be ideal if f were an injection, for it would follow that Dedekind cuts are just the infinite decimal expansions. However this is not true. Suppose that

$$s = .a_1 \ldots a_n a_{n+1} \ldots$$

where $a_n \neq 9$ and for all $k > n$, $a_k = 9$. If

$$t = .a_1 \ldots a_{n-1}(a_n + 1)000\ldots$$

then

$$f(s) = f(t).$$

If neither s nor t are infinite decimal expansions that terminate in repeating 9s, and $s < t$, then there is some n such that $s < t_n$. So the rational number $(s_n + t_n)/2$ is in the Dedekind cut $f(t)$ and not in $f(s)$, so $f(s) \neq f(t)$. Therefore we have proved the following theorem.

THEOREM 8.17. *Let D_0 be the set of infinite decimal expansions for numbers in the unit interval. Let $f : D_0 \to [0,1]$ be defined by*

$$f(.a_1a_2\ldots) = \lim_{n\to\infty} .a_1 \ldots a_n = \sum_{k=1}^{\infty} a_k 10^{-k}.$$

Then f is a surjection. Moreover, two distinct decimal expansions are identified by f iff one of them is of the form $.a_1a_2\ldots a_n 9999\ldots$ with $a_n \neq 9$ and the other is $.a_1a_2\ldots(a_n + 1)000\ldots$.

COROLLARY 8.18. $|\,[0,1]\,| = 2^{\aleph_0}$.

PROOF. By Theorem 8.17, Proposition 6.13, and Theorem 6.9,

$$|\,[0,1]\,| \leq |\,D_0\,| = |\,\ulcorner 10 \urcorner^{\mathbb{N}}\,| = 2^{\aleph_0}.$$

Let $g : \ulcorner 2 \urcorner^{\mathbb{N}^+} \to D_0$ be defined by

$$g(\langle a_n \rangle) = .a_1a_2\ldots$$

and $h : D_0 \to [0,1]$ be defined as in the argument for Theorem 8.17. Then $h \circ g : \ulcorner 2 \urcorner^{\mathbb{N}} \to [0,1]$ is an injection, and so

$$2^{\aleph_0} \leq |\,[0,1]\,|.$$

By the Schröder-Bernstein theorem,

$$|\,[0,1]\,| = 2^{\aleph_0}.$$

$\square$

COROLLARY 8.19. $|\mathbb{R}| = 2^{\aleph_0}$.

8.10 Order-Completeness

We give an argument for the uncountability of $\mathbb{R}$ depending only on its abstract order properties.

DEFINITION.   Order-complete   Let $(X, \leq)$ be a linearly ordered set. It is called order-complete if, whenever A and B are nonempty subsets of X with the property that

$$(\forall\, a \in A)\, (\forall\, b \in B)\quad a \leq b,$$

then there exists c in X such that

$$(\forall\, a \in A)\, (\forall\, b \in B)\quad a \leq c \leq b. \tag{8.20}$$

Note that any order-complete set must have the least upper bound property—if A is any nonempty bounded set, let B be the set of all upper bounds for A, and then c from (8.20) is the (unique) least upper bound for A.

DEFINITION.   Dense   Let $(X, \leq)$ be a linearly ordered set, and $Y \subseteq X$. We say Y is dense in X if

$$(\forall\ a < b \in X)\, (\exists\, y \in Y)\, a < y < b.$$

DEFINITION.   Extension   Let $(X, \leq_X)$ and $(Y, \leq_Y)$ be linearly ordered sets. We say $(Y, \leq_Y)$ is an extension of $(X, \leq_X)$ if $X \subseteq Y$ and, for all x_1, x_2 in X,

$$x_1 \leq_X x_2 \quad\text{iff}\quad x_1 \leq_Y x_2.$$

THEOREM 8.21. *Let $(X, \leq)$ be an extension of $(\mathbb{Q}, \leq)$. If $(X, \leq)$ is order-complete and $\mathbb{Q}$ is dense in X, then X is uncountable.*

PROOF. Suppose that X is a countable order-complete extension of $\mathbb{Q}$ and that $\mathbb{Q}$ is dense in X.

Let the sequence $\langle a_n \mid n \in \mathbb{N} \rangle$ be a bijection from $\mathbb{N}$ to X. Observe that the sequence imposes an ordering on X. Let $\preceq$ be defined on X by

$$(\forall m, n \in \mathbb{N}) \; a_m \preceq a_n \iff m \leq n.$$

That is, for any $x, y \in X$, $x \preceq y$ if x appears in the sequence $\langle a_n \rangle$ before y. Then $\preceq$ is a well-ordering of X.

Given $Y \subseteq X$ and $y_0 \in Y$, we say that y_0 is the $\preceq$-minimal element of Y if

$$(\forall x \in Y) \; y_0 \preceq x.$$

So every subset of X has a $\preceq$-minimal element.

We shall define two subsequences of $\langle a_n \rangle$, called $\langle a_{f(n)} \rangle$ and $\langle a_{g(n)} \rangle$, so that for any $n \in \mathbb{N}$

(1) $f(n+1) > g(n)$

(2) $g(n) > f(n)$

(3) $a_{f(n+1)}$ is the $\preceq$-minimal element of the set

$$\{y \in X \mid a_{f(n)} < y < a_{g(n)}\}$$

(4) $a_{g(n+1)}$ is the $\preceq$-minimal element of the set

$$\{y \in X \mid a_{f(n+1)} < y < a_{g(n)}\}$$

We define the subsequences by recursion using the sequence $\langle a_n \rangle$ to carefully control the construction. This argument is called a back-and-forth argument. Given finite sequences of length N satisfying the properties enumerated above, we define $a_{f(N+1)}$ subject to constraints imposed by $a_{f(N)}$ and $a_{g(N)}$. We then define $a_{g(N+1)}$ subject to constraints imposed by $a_{f(N+1)}$ and $a_{g(N)}$. We then define $a_{f(N+2)}, a_{g(N+2)},$ and so on.

Let $f(0) = 0$. So $a_{f(0)} = a_0$. Let $g(0)$ be the smallest integer n such that $a_0 < a_n$. Note that this is equivalent to defining $g(0)$ so that $a_{g(0)}$ is the $\preceq$-minimal element of X greater than a_0. Assume we have defined finite subsequences $\langle a_{f(n)} \mid n \leq N \rangle$, $\langle a_{g(n)} \mid n \leq N \rangle$ satisfying the order properties listed above. We shall define $a_{f(N+1)}$ and $a_{g(N+1)}$, satisfying the ordering properties listed above. The set X contains the rational numbers, and since $\mathbb{Q}$ is dense in X, there is an element of X, x, such that

$$a_{f(N+1)} < x < g_{(N+1)}.$$

Let $a_{f(N+1)}$ be the $\preceq$-minimal element of X such that

$$a_{f(N)} < a_{f(N+1)} < a_{g(N)}.$$

Since $\preceq$ is a well-ordering of X, $f(N+1)$ is well-defined. We let $a_{g(N+1)}$ be the $\preceq$-minimal element of X such that

$$a_{f(N+1)} < a_{g(N+1)} < a_{g(N)}.$$

By our previous discussion, $g(N+1)$ is well-defined. Observe that for any $m, n \in \mathbb{N}$,

$$a_{f(m)} < a_{g(n)}.$$

Therefore the increasing sequence $\langle a_{f(n)} \mid n \in \mathbb{N} \rangle$ is bounded above, and by Lemma 8.5, the sequence converges to its least upper bound, a.

For any $n \in \mathbb{N}$,

$$a_{f(n)} < a < a_{g(n)}.$$

So a is not a term of either subsequence. We show that a is not a term in the sequence $\langle a_n \rangle$. Suppose by way of contradiction that $a = a_n$ for some $n \in \mathbb{N}$. Since $f(0) = 0$, $n \neq 0$. Let

$$Y = (f[\mathbb{N}] \cup g[\mathbb{N}]) \cap \ulcorner n \urcorner.$$

Then $Y \neq \emptyset$ is finite and has a maximal element.

If the maximal element of Y is $f(0)$, then for every $1 \leq k < n$, we must have $a_k < a_0$. But then $g(0)$ would be n, which contradicts the fact that n is not in the range of g.

If the maximal element of Y is $f(m+1)$ for some m, then $g(m+1) > n$, and

$$f(m+1) < n < g(m+1).$$

However,

$$a_{f(m+1)} < a_n < a_{g(m+1)} < a_{g(m)}.$$

This is impossible since $a_{g(m+1)}$ is the $\preceq$-minimal element of X in the open interval $(a_{f(m+1)}, a_{g(m)})$.

If the maximal element of Y is $g(m)$ for some m, then $f(m+1) > n$ and

$$g(m) < n < f(m+1).$$

However,

$$a_{f(m)} < a_{f(m+1)} < a_n < a_{g(m)}.$$

This is impossible since $a_{f(m+1)}$ is the $\preceq$-minimal element of X in the open interval $(a_{f(m)}, a_{g(m)})$. So a is not a term in the sequence $\langle a_n \rangle$. Therefore there is no bijection from $\mathbb{N}$ to X, and X is uncountable. $\square$

By Exercise 8.20, $\mathbb{Q}$ is dense in $\mathbb{R}$. As the set of real numbers is order-complete by the least upper bound theorem, we get:

COROLLARY 8.22. *The set of real numbers is uncountable.*

THEOREM 8.23. *Let $(X, \leq_X)$ be an order-complete extension of $\mathbb{Q}$ in which $\mathbb{Q}$ is dense and such that X has no maximal or minimal element. Then there is an order-preserving bijection from $\mathbb{R}$ onto X that is the identity on $\mathbb{Q}$.*

PROOF.   Let us define a map $f : \mathbb{R} \to X$. If $q \in \mathbb{Q}$, define $f(q) = q$. If $\alpha \in \mathbb{R} \setminus \mathbb{Q}$, define $f(\alpha)$ to be the least upper bound in X of $\{q \in \mathbb{Q} \mid q \leq \alpha\}$. The function f is well-defined because X has the least upper bound property. It is injective because if $\alpha \neq \beta$, there are rational numbers between α and β.

To show f is onto, suppose $x \in X$. Define $\alpha \in \mathbb{R}$ to be the least upper bound in $\mathbb{R}$ of $\{q \in \mathbb{Q} \mid q \leq_X x\}$. Then $f(\alpha) = x$.

Finally, f is order-preserving because if $\alpha \leq \beta$, then $f(\beta)$ is defined as the least upper bound of a superset of the set whose least upper bound is $f(\alpha)$, and so $f(\alpha) \leq_X f(\beta)$. $\qquad\qquad$ □

REMARK. What happens if we drop the requirement that X have no maximal or minimal element?

8.11 Exercises

EXERCISE 8.1. Let S be the successor function defined on p. 225. Prove that

$$S(\emptyset) \neq \emptyset.$$

Prove that for any set X,

$$S(X) \neq X.$$

EXERCISE 8.2. Prove that no proper subset of $\mathbf{N}$ (see equation 8.1) is inductive.

EXERCISE 8.3. Let $\mathcal{F} = \{X_\alpha \mid \alpha \in Y\}$ be a family of inductive sets indexed by Y. Prove that

$$\bigcap_{\alpha \in Y} X_\alpha$$

is inductive.

EXERCISE 8.4. Prove that addition and multiplication in $\mathbb{N}$ (as formally defined in Section 8.1) are associative, commutative, and distributive.

EXERCISE 8.5. Prove that the relation $\leq$ defined on $\mathbb{N}$ in Section 8.1 is a linear ordering of $\mathbb{N}$.

EXERCISE 8.6. Prove that addition and multiplication in $\mathbb{Z}$ (as formally defined in Section 8.2) are associative, commutative, and distributive.

EXERCISE 8.7. Prove that the relation $\leq$ defined on $\mathbb{Z}$ in Section 8.2 is a linear ordering of $\mathbb{Z}$.

EXERCISE 8.8. Prove that $\leq$ is a well-ordering of $\mathbb{N}$ but not of $\mathbb{Z}$ (using the formal definition of the relation).

EXERCISE 8.9. Prove that addition and multiplication in $\mathbb{Z}$ and the relation $\leq$ on $\mathbb{Z}$ extends the operations and relation on $\mathbb{N}$. Let $I : \mathbb{N} \to \mathbb{Z}$ be defined by

$$I(n) \;=\; [\langle n, 0 \rangle].$$

Prove that I is an injection and that for all $m, n \in \mathbb{N}$,

$$I(m + n) \;=\; I(m) + I(n), \tag{8.24}$$

$$I(m \cdot n) \;=\; I(m) \cdot I(n), \tag{8.25}$$

and

$$m \leq n \;\Rightarrow\; I(m) \leq I(n). \tag{8.26}$$

Note that the operations on the left-hand sides of equations 8.24 and 8.25 are defined in $\mathbb{N}$ and on the right-hand side are defined in $\mathbb{Z}$. Similarly, the antecedent of statement 8.26 is defined in $\mathbb{N}$, and the consequence is defined in $\mathbb{Z}$.

EXERCISE 8.10. Prove that addition and multiplication in $\mathbb{Q}$ (as formally defined in Section 8.3) are associative, commutative, and distributive.

EXERCISE 8.11. Prove that the relation $\leq$ defined on $\mathbb{Q}$ in Section 8.3 is a linear ordering of $\mathbb{Q}$.

EXERCISE 8.12. Prove that addition and multiplication on $\mathbb{Q}$ and the relation $\leq$ on $\mathbb{Q}$ extends the operations and relation on $\mathbb{Q}$. Let $I : \mathbb{Z} \to \mathbb{Q}$ be defined by

$$I(a) \;=\; [\langle a, 1 \rangle].$$

Prove that I is an injection and that for all $a, b \in \mathbb{Z}$,

$$I(a + b) \;=\; I(a) + I(b), \tag{8.27}$$

$$I(a \cdot b) \;=\; I(a) \cdot I(b), \tag{8.28}$$

and

$$a \leq b \;\Rightarrow\; I(a) \leq I(b). \tag{8.29}$$

Note that the operations on the left-hand sides of equations 8.27 and 8.28 are defined in $\mathbb{Z}$ and on the right-hand side are defined in $\mathbb{Q}$. Similarly, the antecedent of statement 8.29 is defined in $\mathbb{Z}$, and the consequence is defined in $\mathbb{Q}$.

EXERCISE 8.13. Prove that every nonzero element of $\mathbb{Q}$ has a multiplicative inverse in $\mathbb{Q}$.

EXERCISE 8.14. Prove statements $(1), (2)$ and (3) in Section 8.4.

EXERCISE 8.15. Prove Theorem 8.2.

EXERCISE 8.16. Let $X \subseteq \mathbb{R}$, $Y \subseteq \mathbb{R}$, and let every element of X be less than every element of Y. Prove that there is $a \in \mathbb{R}$ satisfying

$$(\forall x \in X)(\forall y \in Y)\; x \leq a \leq y.$$

EXERCISE 8.17. Let $X \subseteq \mathbb{R}$ be bounded above. Prove that the least upper bound of X is unique.

EXERCISE 8.18. Let $X \subseteq \mathbb{R}$ be bounded below. Prove that X has a greatest lower bound.

EXERCISE 8.19. Only the special case of the Bolzano-Weierstrass Theorem (Theorem 8.6) was proved (where $[b, c]$ is the closed unit interval, $[0, 1]$). Generalize the proof to arbitrary $b, c \in \mathbb{R}$ where $b \leq c$.

EXERCISE 8.20. Let $X \subseteq \mathbb{R}$. We say that X is dense in $\mathbb{R}$ if, given any $a, b \in \mathbb{R}$ with $a < b$, there is $x \in X$ such that

$$a \leq x \leq b.$$

a) Prove that $\mathbb{Q}$ is dense in $\mathbb{R}$.

b) Prove that $\mathbb{R} \setminus \mathbb{Q}$ is dense in $\mathbb{R}$.

EXERCISE 8.21. Let $\langle a_n \rangle$ be an injective sequence. What is the cardinality of the set of all subsequences of $\langle a_n \rangle$? What can you say about the set of subsequences of a noninjective sequence?

EXERCISE 8.22. Let s be an infinite decimal expansion, and for any $n \in \mathbb{N}^+$, let s_n be the truncation of s to the nth decimal place. Prove that the sequence $\langle s_n \rangle$ is a Cauchy sequence.

EXERCISE 8.23. Let $\langle a_n \rangle$ be a convergent sequence and $\langle a_{f(n)} \rangle$ be a subsequence of $\langle a_n \rangle$. Prove that

$$\lim_{n \to \infty} a_n = \lim_{n \to \infty} a_{f(n)}.$$

EXERCISE 8.24. Prove the following generalization of the triangle inequality: if the series $\sum_{n=0}^{\infty} a_n$ converges, then

$$\left| \sum_{n=0}^{\infty} a_n \right| \leq \sum_{n=0}^{\infty} | a_n | .$$

EXERCISE 8.25. Let f be a real function continuous at a, and let $\langle a_n \rangle$ be a sequence converging to a. Prove that

$$\lim_{n \to \infty} f(a_n) = f(a).$$

EXERCISE 8.26. Give an example of a continuous function on an open interval that achieves its extreme values on the interval. Give an example of a continuous function defined on an open interval that does not achieve its extreme values on the interval.

EXERCISE 8.27. Complete the proof of Theorem 8.12—that is, prove the result for $f(c)$ a minimum value of f on (a, b).

EXERCISE 8.28. Prove Corollary 8.15.

EXERCISE 8.29. Prove that any continuous injective real function on an interval is monotonic on that interval.

EXERCISE 8.30. Prove that there is no continuous bijection from $(0, 1)$ to $[0, 1]$.

EXERCISE 8.31. Prove that every polynomial in $\mathbb{R}[x]$ of odd degree has at least one real root.

EXERCISE 8.32. Prove that if you have a square table with legs of equal length and a continuous floor, you can always rotate the table so that all four legs are simultaneously in contact with the floor. (Hint: Apply the Intermediate Value theorem to an appropriately chosen function.) This is one of the earliest applications of mathematics to coffeehouses.

EXERCISE 8.33. The proof of Proposition 8.16 requires that nonzero real numbers have reciprocals (and hence quotients of real numbers are well-defined). Prove that nonzero real numbers have reciprocals.

EXERCISE 8.34. Show that there are exactly four order-complete extensions of $\mathbb{Q}$ in which $\mathbb{Q}$ is dense.

CHAPTER 9

Complex Numbers

9.1 Cubics

How does one find the roots of a cubic polynomial? The Babylonians knew the quadratic formula in the second millennium BC, but a formula for the cubic was only found in the sixteenth century. The history of the discovery is complicated, but most of the credit should go to Nicolo Tartaglia. The solution was published in 1545 in Girolomo Cardano's very influential book *Artis magnae sive de regulis algebraicis liber unus*. Formula 9.7 is known today as the Tartaglia-Cardano formula. For a historical account, see, for example, [6].

Consider a cubic polynomial in $\mathbb{R}[x]$

$$p(x) = a_3 x^3 + a_2 x^2 + a_1 x + a_0. \tag{9.1}$$

If we want to find the roots, there is no loss of generality in assuming that $a_3 = 1$, since the zeros of p are the same as the zeros of $\frac{1}{a_3} p$.

The second simplification is that we can assume $a_2 = 0$. Indeed, make the change of variable

$$x = y - \beta$$

for some β to be chosen later. Then

$$
\begin{aligned}
p(x) &= x^3 + a_2 x^2 + a_1 x + a_0 \\
&= (y - \beta)^3 + a_2 (y - \beta)^2 + a_1 (y - \beta) + a_0 \\
&= y^3 + [a_2 - 3\beta] y^2 + [a_1 - 2a_2\beta + 3\beta^2] y + [a_0 - a_1\beta + a_2\beta^2 - \beta^3] \\
&=: q(y).
\end{aligned}
$$

Choose $\beta = a_2/3$. Then the coefficient of y^2 in $q(y)$ vanishes. Suppose you can find the roots of q; call them $\alpha_1, \alpha_2, \alpha_3$. Then the roots of the original polynomial p are $\alpha_1 - \beta$, $\alpha_2 - \beta$, and $\alpha_3 - \beta$.

Therefore it is sufficient to find a formula for the roots of a cubic in which the quadratic term vanishes. This is called a *reduced cubic*. As there are now only two coefficients left, we shall drop the subscripts and write our reduced cubic as

$$q(x) = x^3 + ax + b. \tag{9.2}$$

The key idea is to make another, more ingenious, substitution. Let us introduce a new variable w, related to x by

$$x = w + \frac{c}{w}, \tag{9.3}$$

where c is a constant that we shall choose later. Then

$$
\begin{aligned}
q(x) &= \left(w + \frac{c}{w}\right)^3 + a\left(w + \frac{c}{w}\right) + b \\
&= w^3 + [3c + a]w + [3c^2 + ac]\frac{1}{w} + c^3\frac{1}{w^3} + b. \tag{9.4}
\end{aligned}
$$

Choose

$$c = -\frac{a}{3},$$

so both the coefficient of w and $1/w$ in (9.4) vanish. Then finding x so that $q(x) = 0$ is the same as finding w so that

$$
\begin{aligned}
w^3 + \frac{c^3}{w^3} + b &= 0 \\
\Longleftrightarrow w^6 + bw^3 + c^3 &= 0. \tag{9.5}
\end{aligned}
$$

Equation (9.5) is of degree 6, which seems worse than the original cubic; but so many terms vanish that it is actually a quadratic equation in w^3. Therefore it can solved by the quadratic formula:

$$w^3 = \frac{-b \pm \sqrt{b^2 - 4c^3}}{2}. \tag{9.6}$$

Knowing w, we can recover x by

$$x = w + \frac{c}{w} = w - \frac{a}{3w}.$$

So we arrive at the Tartaglia-Cardano formula for the roots of the reduced cubic (9.2):

$$x = \left[\frac{-b \pm \sqrt{b^2 + \frac{4a^3}{27}}}{2}\right]^{1/3} - \frac{a}{3\left[\frac{-b \pm \sqrt{b^2 + \frac{4a^3}{27}}}{2}\right]^{1/3}}. \qquad (9.7)$$

How does the formula work in practice?

EXAMPLE 9.8. Let $p(x) = x^3 - 3x + 2$. Then $c = 1$, and (9.6) says $w^3 = -1$. Therefore $w = -1$, and so $x = -2$ is a root. Therefore, by Lemma 4.13, $(x + 2)$ is a factor of p. Factoring, we get

$$x^3 - 3x + 2 = (x + 2)(x^2 - 2x + 1).$$

The last term factors as $(x - 1)^2$, so we conclude that the roots are $-2, 1, 1$.

In Example 9.8, the formula worked but only gave us one of the roots. Consider the next example.

EXAMPLE 9.9. Let

$$p(x) = x^3 - 3x + 1. \qquad (9.10)$$

Then $c = 1$, and

$$w^3 = \frac{-1 \pm \sqrt{-3}}{2}. \qquad (9.11)$$

Now we have a worse problem: w^3 involves the square root of a negative number, and even if we make sense of that, we then have to extract a cube root. Is this analagous to trying to solve the quadratic equation

$$q(x) := x^2 + x + 1 = 0?$$

The quadratic formula again gives the right-hand side of (9.11), and we explain this by saying that in fact q has no real roots. Indeed, graphing shows that q looks like Figure 9.1.

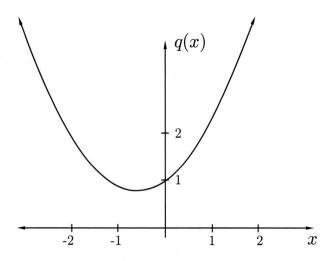

FIGURE 9.1 Plot of $q(x) = x^2 + x + 1$

But this cannot be the case for p. Indeed,

$$
\begin{aligned}
p(-2) &= -1 < 0 \\
p(0) &= 1 > 0 \\
p(1) &= -1 < 0 \\
p(2) &= 3 > 0.
\end{aligned}
$$

Therefore, by Theorem 8.10 (the Intermediate Value theorem), p must have a root in each of the intervals $(-2, 0)$, $(0, 1)$, and $(1, 2)$. As p can have at most three roots by Theorem 4.10, it must therefore have exactly three roots. A graph of p looks like Figure 9.2.

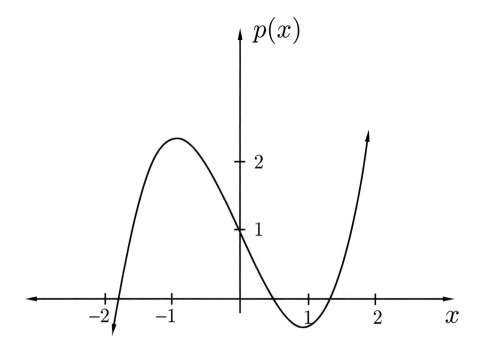

FIGURE 9.2 Plot of $p(x) = x^3 - 3x + 1$

It turns out that one can find the roots of p in Example 9.9 by correctly interpreting the Tartaglia-Cardano formula. We shall come back to this example in Section 9.3, after we develop the necessary ideas. The big idea is to introduce the notion of a *complex number*.

9.2 Complex Numbers

DEFINITION. Complex number A complex number is an expression of the form $a + ib$, where a and b are real numbers.

For the moment, you can think of the i in $a + ib$ as a formal symbol or a placeholder. Later, we shall see that it has another interpretation.

NOTATION. $\mathbb{C}$ We shall let $\mathbb{C}$ denote the set of all complex numbers:

$$\mathbb{C} = \{a + ib : a, b \in \mathbb{R}\}.$$

As a set, one can identify $\mathbb{C}$ with $\mathbb{R}^2$ in the obvious way. This allows us to define addition; what is not so obvious is that there is also a good definition for multiplication.

DEFINITION. Let $a + ib$ and $c + id$ be complex numbers. Then their sum and product are defined by

$$(a + ib) + (c + id) \quad = \quad (a + c) + i(b + d) \tag{9.12}$$

$$(a + ib) \times (c + id) \quad = \quad (ac - bd) + i(ad + bc). \tag{9.13}$$

The formula for the sum (9.12) is just what you would get if you identified the complex number $a + ib$ with the vector (a, b) in $\mathbb{R}^2$ and used vector addition. The product is more subtle. If you multiply out the left-hand side of (9.13), you get

$$ac + i(ad + bc) + i^2 bd.$$

One arrives at the right-hand side of (9.13) by *defining*

$$i^2 = -1. \tag{9.14}$$

So i is the square root of -1; that is, it is an algebraic quantity we introduce that is defined to have the property that its square is -1. Obviously this precludes i from being a real number.

In essence we have continued the program of defining number systems that we began in Chapter 8. Addition and multiplication of complex numbers have been defined by algebraic operations on $\mathbb{R} \times \mathbb{R}$. Since algebraic operations on the real numbers were defined set-theoretically, we have thereby defined algebraic operations on $\mathbb{C}$ by set operations. Unlike the other numbers systems we have defined, we do not define a linear ordering of $\mathbb{C}$. It is not generally useful to think of complex numbers on a number line. However it is very useful to think of complex numbers as points in the plane $\mathbb{R}^2$ and to describe them in polar coordinates.

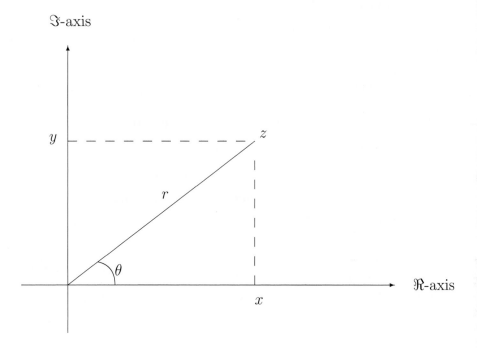

FIGURE 9.3 Polar coordinates

As usual, the point that has Cartesian coordinates (x, y) has polar coordinates (r, θ), where they are related by

$$r = \sqrt{x^2 + y^2} \qquad \tan(\theta) = y/x$$
$$x = r \cos \theta \qquad y = r \sin \theta.$$

So the complex number $z = x + iy$ can also be written as

$$z = r(\cos \theta + i \sin \theta). \qquad (9.15)$$

The form (9.15) is so widely used that there is a special notation for it.

NOTATION.   Cis

$$\mathrm{Cis}(\theta) \ := \ \cos\theta + i\sin\theta.$$

DEFINITION. For the complex number $z \ = \ x + iy \ = \ r\mathrm{Cis}(\theta)$, we have the following:

$\mathfrak{R}(z)$ x is called the *real part* of z, written $\mathfrak{R}(z)$.

$\mathfrak{I}(z)$ y is called the *imaginary part* of z, written $\mathfrak{I}(z)$.

$|z|$ r is called the *modulus* of z, or *absolute value* of z, written $|z|$.

$\arg(z)$ θ is called the *argument* of z, written $\arg(z)$.

$\bar{z}$ The number $x - iy$ is called the *conjugate* of z, written $\bar{z}$.

REMARK. There is an important point to bear in mind about the argument: it is only unique up to addition of multiples of 2π. In other words, if θ_0 is an argument of the complex number z, then so are all the numbers $\{\theta_0 + 2k\pi \ : \ k \in \mathbb{Z}\}$.

Addition is easiest in Cartesian coordinates: add the real and imaginary parts. Multiplication is easiest in polar coordinates: multiply the moduli and add the arguments.

PROPOSITION 9.16. *Let* $z_1 = r_1\mathrm{Cis}(\theta_1)$ *and* $z_2 = r_2\mathrm{Cis}(\theta_2)$. *Then*

$$z_1 z_2 \ = \ r_1 r_2 \mathrm{Cis}(\theta_1 + \theta_2).$$

PROOF. Multiplying out, we get

$$z_1 z_2 \ = \ r_1 r_2 [\cos\theta_1 \cos\theta_2 - \sin\theta_1 \sin\theta_2$$
$$+ \ i \, (\cos\theta_1 \sin\theta_2 + \cos\theta_2 \sin\theta_1)].$$

The result follows from the trigonometric identities for the cosine and sine of the sum of two angles.   □

A consequence of Proposition 9.16 is the following formula for raising a complex number to a power, called De Moivre's theorem.

THEOREM 9.17. *De Moivre's theorem Let $z = r\text{Cis}(\theta)$ be a nonzero complex number, and let $n \in \mathbb{Z}$. Then*

$$z^n = r^n \text{Cis}(n\theta). \tag{9.18}$$

PROOF. If $n \geq 0$, then (9.18) can be proved by induction from Proposition 9.16. For n negative, it is enough to observe that by Proposition 9.16

$$[r\text{Cis}(\theta)]\left[r^{-1}\text{Cis}(-\theta)\right] = 1\text{Cis}(0) = 1.$$

$\square$

We can now prove that every nonzero complex number has *exactly* n distinct n^{th} roots.

THEOREM 9.19. *Let $z = r\text{Cis}(\theta)$ be a nonzero complex number, and let n be an integer greater than 1. Then there are exactly n complex numbers w satisfying the equation $w^n = z$. They are*

$$\left\{ r^{1/n}\text{Cis}\left(\frac{\theta}{n} + \frac{2k\pi}{n}\right) : k = 0, 1, \ldots, n-1 \right\}. \tag{9.20}$$

PROOF. Suppose $w = \rho\text{Cis}(\phi)$ is an nth root of z. Then by De Moivre's theorem, $\rho^n = r$ and $n\phi$ is an argument of z. As ρ must be a positive real number, it is the unique positive n^{th} root of r. The number $n\phi$ can be *any* argument of z, so we have

$$n\phi = \theta + 2k\pi, \quad k \in \mathbb{Z}.$$

So ϕ can have the form

$$\frac{\theta}{n} + \frac{2k\pi}{n}$$

for any integer k. However, different ϕ's will give rise to the same complex number w if they differ by a multiple of 2π. So there are exactly n different w's that are n^{th} roots of z. $\square$

EXAMPLE 9.21. What does Theorem 9.19 tell us are the square roots of -1? We let $r = 1$ and $\theta = \pi$, and we get the square roots $\text{Cis}(\pi/2) = i$ and $\text{Cis}(-\pi/2) = -i$.

EXAMPLE 9.22. Find the cube roots of 1.

In the notation of Theorem 9.19, $r = 1$ and $\theta = 0$. Therefore the cube roots are

$$1$$

$$\omega = \operatorname{Cis}(2\pi/3) = -\frac{1}{2} + i\frac{\sqrt{3}}{2}$$

$$\omega^2 = \operatorname{Cis}(4\pi/3) = -\frac{1}{2} - i\frac{\sqrt{3}}{2}.$$

The number ω is called a *primitive cube root of unity* because all the cube roots are obtained as $\omega, \omega^2, \omega^3$.

DEFINITION. Primitive root of unity A primitive n^{th} root of unity is a number ω such that $\{1, \omega, \omega^2, \ldots, \omega^{n-1}\}$ constitute all the nth roots of 1.

PROPOSITION 9.23. *Let z be a complex number, and w_0 be some n^{th} root of z. Let ω be a primitive n^{th} root of unity. Then all the nth roots of z are $\{w_0, \omega w_0, \omega^2 w_0, \ldots, \omega^{n-1} w_0\}$.*

9.3 Tartaglia-Cardano Revisited

Let us consider again Example 9.9. We wanted to find the cube roots of

$$\zeta_\pm = \frac{-1 \pm \sqrt{-3}}{2}.$$

If we take the plus sign, we get

$$\zeta_+ = \operatorname{Cis}(2\pi/3),$$

and if we take the minus sign, we get

$$\zeta_- = \operatorname{Cis}(4\pi/3).$$

So ζ_+ has three roots, namely

$$\left\{\operatorname{Cis}\left(\frac{2\pi}{9} + \frac{2k\pi}{3}\right) : k = 0, 1, 2\right\},$$

and ζ_- has three roots, namely

$$\{\operatorname{Cis}\left(\frac{4\pi}{9} + \frac{2k\pi}{3}\right) : k = 0, 1, 2\}.$$

Knowing w, we want to find x, which for Example 9.9 is given by $w + 1/w$. For any number w that can be written as $\operatorname{Cis}(\theta)$ (i.e., any complex number of modulus 1), we have

$$w + \frac{1}{w} = \cos\theta + i\sin\theta + \cos(-\theta) + i\sin(-\theta)$$
$$= 2\cos\theta.$$

Therefore the roots of the polynomial given in (9.10) are

$$\left\{2\cos\frac{2\pi}{9}, 2\cos\frac{8\pi}{9}, 2\cos\frac{14\pi}{9}, 2\cos\frac{4\pi}{9}, 2\cos\frac{10\pi}{9}, 2\cos\frac{16\pi}{9}\right\}. \quad (9.24)$$

Are these six different roots? Theorem 4.10 says that p can have at most three different roots. As $\cos(\theta) = \cos(2\pi - \theta)$, we see our set (9.24) may be written as

$$\{2\cos\frac{2\pi}{9}, 2\cos\frac{4\pi}{9}, 2\cos\frac{8\pi}{9}\}. \quad (9.25)$$

It turns out that the Tartaglia-Cardano formula (9.7) does give all three roots of the cubic, and moreover it does not matter whether one chooses the plus or minus sign, as long as one calculates all three cube roots of (9.6) for some choice of sign. We shall use $\mathbb{C}[z]$ to denote the set of polynomials in z with coefficients from $\mathbb{C}$.

THEOREM 9.26. *Consider the polynomial*

$$p(z) = z^3 + az + b \quad (9.27)$$

in $\mathbb{C}[z]$, and assume $a \neq 0$. Let $c = -a/3$, and let ζ be

$$\zeta = \frac{-b + \sqrt{b^2 - 4c^3}}{2}. \quad (9.28)$$

Let w_1, w_2, and w_3 be the three distinct cube roots of ζ. For each w_i, define z_i by

$$z_i = w_i + \frac{c}{w_i}.$$

Then

$$p(z) = (z - z_1)(z - z_2)(z - z_3). \tag{9.29}$$

REMARK. It will follow from the proof that it does not matter which square root of $b^2 - 4c^3$ one chooses in (9.28).

PROOF. If p is given by (9.29), then

$$p(z) = z^3 - (z_1 + z_2 + z_3)z^2 + (z_1z_2 + z_2z_3 + z_3z_1)z - (z_1z_2z_3). \tag{9.30}$$

We must show that the coefficients in (9.30) match those in (9.27). By Proposition 9.23, we can assume

$$w_1 = \omega w_3, \qquad w_2 = \omega^2 w_3$$

where $\omega = -\frac{1}{2} + i\frac{\sqrt{3}}{2}$ is a primitive cube root of unity. In the following calculations, we use the facts that $\omega^2 = 1/\omega$ and $1 + \omega + \omega^2 = 0$. (Why are these true?) Notice that $w_3 \neq 0$, as that would force $c = 0$.

The coefficient of z^2 in (9.30) is

$$
\begin{aligned}
-(z_1 + z_2 + z_3) &= -w_3(\omega + \omega^2 + 1) - \frac{1}{w_3}(\omega^2 + \omega + 1) \\
&= 0.
\end{aligned}
$$

The coefficient of z is

$$
\begin{aligned}
z_1z_2 + z_2z_3 + z_3z_1 &= \left(\omega w_3 + c\omega^2 \frac{1}{w_3}\right)\left(\omega^2 w_3 + c\omega \frac{1}{w_3}\right) \\
&\quad + \left(\omega^2 w_3 + c\omega \frac{1}{w_3}\right)\left(w_3 + c \frac{1}{w_3}\right) \\
&\quad + \left(w_3 + c \frac{1}{w_3}\right)\left(\omega w_3 + c\omega^2 \frac{1}{w_3}\right) \\
&= w_3^2(1 + \omega^2 + \omega) + 3c(\omega + \omega^2) + \frac{c^2}{w_3^2}(1 + \omega + \omega^2) \\
&= -3c \\
&= a.
\end{aligned}
$$

The constant term in (9.30) is

$$
\begin{aligned}
-z_1 z_2 z_3 &= -\left(\omega w_3 + c\omega^2 \frac{1}{w_3}\right)\left(\omega^2 w_3 + c\omega \frac{1}{w_3}\right)\left(w_3 + \frac{1}{w_3}\right) \\
&= -w_3^3 - cw_3(1 + \omega^2 + \omega) - \frac{c^2}{\frac{1}{w_3}}(\omega + 1 + \omega^2) - \frac{c^3}{w_3^3} \\
&= -\zeta - \frac{c^3}{\zeta} \\
&= -\frac{-b + \sqrt{b^2 - 4c^3}}{2} - \frac{2c^3}{-b + \sqrt{b^2 - 4c^3}} \\
&= \frac{-b^2 + 2b\sqrt{b^2 - 4c^3} - (b^2 - 4c^3) - 4c^3}{2(-b + \sqrt{b^2 - 4c^3})} \\
&= \frac{b(-b + \sqrt{b^2 - 4c^3})}{-b + \sqrt{b^2 - 4c^3}} \\
&= b.
\end{aligned}
$$

Therefore, all the coefficients of (9.27) and (9.29) match, so they are the same polynomial. □

The Tartaglia-Cardano formula therefore gives all three roots to a reduced cubic polynomial p with complex coefficients (repeated roots can occur). If the coefficients a and b are real, we know from the Intermediate Value theorem that at least one of the three roots of p will be real (see Exercise 8.31). As Example 9.9 shows, however, it may still be necessary to take the cube root of a complex ζ to obtain the real roots of a real cubic. This realization was what led to the acceptance of complex numbers as useful objects rather than as a bizarre fantasy.

9.4 Fundamental Theorem of Algebra

Algebra over the complex numbers is in many ways easier than over the real numbers. The reason is that a polynomial of degree N in $\mathbb{C}[z]$ has *exactly* N zeros, counting multiplicity. This is called the Fundamental

Theorem of Algebra. To prove it, we must establish some preliminary results.

9.4.1 Some Analysis.

DEFINITION. We say that a sequence $\langle z_n = x_n + iy_n \rangle$ of complex numbers *converges* to the number $z = x + iy$ iff $\langle x_n \rangle$ converges to x and $\langle y_n \rangle$ converges to y. We say the sequence is Cauchy iff both $\langle x_n \rangle$ and $\langle y_n \rangle$ are Cauchy.

REMARK. This is the same as saying that $\langle z_n \rangle$ converges to z iff $|z - z_n|$ tends to zero, and that $\langle z_n \rangle$ is Cauchy iff

$$(\forall \varepsilon > 0)\,(\exists N)\,(\forall m, n > N)\,|z_m - z_n| < \varepsilon.$$

DEFINITION. Let $G \subseteq \mathbb{C}$. We say a function $f : G \to \mathbb{C}$ is continuous on G if, whenever $\langle z_n \rangle$ is a sequence in G that converges to some value z_∞ in G, then $\langle f(z_n) \rangle$ converges to $f(z_\infty)$.

PROPOSITION 9.31. *Polynomials are continuous functions on* $\mathbb{C}$.

PROOF. Repeat the proof of Proposition 5.20 with complex numbers instead of real numbers. □

DEFINITION. Closed rectangle A *closed rectangle* is a set of the form $\{z \in \mathbb{C} \mid a \leq \Re(z) \leq b,\ c \leq \Im(z) \leq d\}$ for some real numbers $a \leq b$ and $c \leq d$.

We would like a version of the Extreme Value theorem, but it is not clear how the minimum and maximum values of a complex-valued function should be defined. However, our definition of continuity makes sense even if the range of f is contained in $\mathbb{R}$, and every complex-valued continuous function g has three naturally associated real-valued continuous functions, viz. $\Re(g), \Im(g)$ and $|g|$.

THEOREM 9.32. *Let R be a closed rectangle in $\mathbb{C}$, and $f : R \to \mathbb{R}$ a continuous function. Then f attains its maximum and its minimum.*

PROOF. Let $R = \{z \in \mathbb{C} \mid a \le \Re(z) \le b, \ c \le \Im(z) \le d\}$. Let $\langle z_n = x_n + iy_n \rangle$ be a sequence of points such that $f(z_n)$ tends to either the least upper bound of the range of f, if this exists, or let $f(z_n) > n$ for all n, if the range is not bounded above. By Theorem 8.6 (the Bolzano-Weierstrass theorem), there is some subsequence for which the real parts converge to some number x_∞ in $[a, b]$. By the Bolzano-Weierstrass theorem again, some subsequence of this subsequence has the property that the imaginary parts also converge, to some point y_∞ in $[c, d]$. So, replacing the original sequence by this subsequence of the subsequence, we can assume that z_n converges to the point $z_\infty = x_\infty + iy_\infty \in R$. By continuity, $f(z_\infty) = \lim_{n \to \infty} f(z_n)$. If the original sequence were unbounded, then $f(z_n) > n$ in the subsequence. This is impossible since the sequence $\langle f(z_n) \rangle$ converges to $f(z_\infty)$. Therefore the subsequence is bounded and $f(z_\infty)$ must be the least upper bound of the range of f. Therefore $f(z_\infty)$ is the maximum of f over R.

A similar argument shows that the minimum is also attained.   □

REMARK. The previous theorem can be improved to show that a continuous real-valued function on a closed bounded set in $\mathbb{C}$ attains its extrema. A set F is *closed* if whenever a sequence of points $\langle z_n \rangle$ converges to some complex number z_∞, then z_∞ is in F. A set is *bounded* if it is contained in some rectangle.

We need one more geometric fact.

LEMMA 9.33. *Triangle inequality Let z_1, z_2 be complex numbers. Then*

$$|z_1 + z_2| \le |z_1| + |z_2|$$

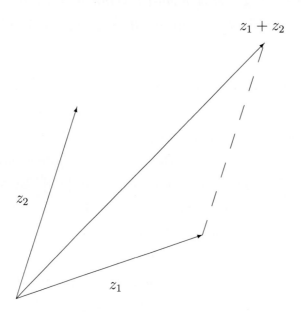

FIGURE 9.4 Triangle inequality

PROOF. Write $z_1 = r_1 \mathrm{Cis}(\theta_1)$ and $z_2 = r_2 \mathrm{Cis}(\theta_2)$. Then

$$|r_1 \mathrm{Cis}(\theta_1) + r_2 \mathrm{Cis}(\theta_2)|$$

$$= \left[(r_1 \cos\theta_1 + r_2 \cos\theta_2)^2 + (r_1 \sin\theta_1 + r_2 \sin\theta_2)^2 \right]^{1/2}$$

$$= \left[r_1^2 + r_2^2 + 2r_1 r_2 (\cos\theta_1 \cos\theta_2 + \sin\theta_1 \sin\theta_2) \right]^{1/2}$$

$$= \left[r_1^2 + r_2^2 + 2r_1 r_2 \cos(\theta_1 - \theta_2) \right]^{1/2}$$

$$\leq \left[r_1^2 + r_2^2 + 2r_1 r_2 \right]^{1/2}$$

$$= r_1 + r_2.$$

$\square$

COROLLARY 9.34. *Let* $z_1, \ldots, z_n \in \mathbb{C}$. *Then*

$$|z_1 + \cdots + z_n| \leq |z_1| + \cdots + |z_n|.$$

9.4.2 The Proof of the Fundamental Theorem of Algebra.

First we observe that finding roots and finding factors are closely related.

LEMMA 9.35. *Let p be a polynomial of degree $N \geq 1$ in $\mathbb{C}[z]$. A complex number, c, is a root of p iff*

$$p(z) = (z - c)q(z),$$

where q is a polynomial of degree $N - 1$.

PROOF. Repeat the proof of Lemma 4.13 with real numbers replaced by complex numbers.   □

Now we prove D'Alembert's lemma, which states that the modulus of a polynomial cannot have a local minimum except at a root.

LEMMA 9.36. D'Alembert's lemma *Let $p \in \mathbb{C}[z]$ and $\alpha \in \mathbb{C}$. If $p(\alpha) \neq 0$, then*

$$(\forall \varepsilon > 0) \, (\exists \zeta) \, [\, |\zeta - \alpha| < \varepsilon] \wedge [\, |p(\zeta)| < |p(\alpha)| \,]. \qquad (9.37)$$

PROOF. Fix α, not a root of p. Write p as

$$p(z) = \sum_{k=0}^{N} a_k (z - \alpha)^k,$$

where neither a_0 nor a_N are 0. Let

$$m = \min\{j \in \mathbb{N}^+ \mid a_j \neq 0\}.$$

So

$$p(z) = a_0 + a_m(z - \alpha)^m + \cdots + a_N(z - \alpha)^N. \qquad (9.38)$$

Let $a_0 = r_0 \mathrm{Cis}(\theta_0)$ and $a_m = r_m \mathrm{Cis}(\theta_m)$. We will choose ζ of the form

$$\zeta = \alpha + \rho \mathrm{Cis}(\phi)$$

in such a way as to get some cancellation in the first two terms of
(9.38). So, let

$$\phi = \frac{\theta_0 + \pi - \theta_m}{m}.$$

Then

$$a_0 + a_m(\zeta - \alpha)^m = r_0\text{Cis}(\theta_0) - r_m\rho^m\text{Cis}(\theta_0).$$

It remains to show that, for ρ small enough, we can ignore all the
higher-order terms. Note that if $\rho < 1$, we have

$$|a_{m+1}(\zeta - \alpha)^{m+1} + \cdots + a_N(\zeta - \alpha)^N|$$
$$\leq |a_{m+1}(\zeta - \alpha)^{m+1}| + \cdots + |a_N(\zeta - \alpha)^N|$$
$$= |a_{m+1}|\rho^{m+1} + \cdots + |a_N|\rho^N$$
$$\leq \rho^{m+1}[|a_{m+1}| + \cdots + |a_N|]$$
$$=: C\rho^{m+1}.$$

Choose ρ so that $r_m\rho^m < r_0$. Then

$$p(\zeta) = (r_0 - r_m\rho^m)\text{Cis}(\theta_0) + a_{m+1}(\zeta - \alpha)^{m+1} + \cdots + a_N(\zeta - \alpha)^N,$$

so

$$|p(\zeta)| \leq r_0 - r_m\rho^m + C\rho^{m+1}. \tag{9.39}$$

If $\rho < r_m/C$, the right-hand side of (9.39) is smaller than r_0.

So we conclude that by taking

$$\rho = \frac{1}{2}\min\left(1, \frac{r_m}{C}, \left[\frac{r_0}{r_m}\right]^{1/m}, \varepsilon\right)$$

then

$$\zeta = \rho\text{Cis}\left(\frac{\theta_0 + \pi - \theta_m}{m}\right)$$

satisfies the conclusion of the lemma.   □

THEOREM 9.40.   *Fundamental Theorem of Algebra* Let $p \in \mathbb{C}[z]$
be a polynomial of degree $N \geq 1$. Then p can be factored as

$$p(z) = c(z - \alpha_1)\ldots(z - \alpha_N) \tag{9.41}$$

for complex numbers $c, \alpha_1, \ldots, \alpha_N$. Moreover, the factoring is unique up to order.

PROOF. (i) Show that p has at least one root.

Let $p(z) = \sum_{k=0}^{N} a_k z^k$, with $a_N \neq 0$. Let S be the closed square $\{z \in \mathbb{C} \mid -L \leq \Re(z) \leq L, \, -L \leq \Im(z) \leq L\}$, where L is some (large) number to be chosen later.

If $|z| = R$ then

$$\left| \sum_{k=0}^{N-1} a_k z^k \right| \leq \sum_{k=0}^{N-1} |a_k| R^k.$$

Choose L_0 so that if $R \geq L_0$, then

$$\sum_{k=0}^{N-1} |a_k| R^k \leq \frac{1}{2} |a_N| R^N.$$

Then if $L \geq L_0$ and z is outside S, we have

$$
\begin{aligned}
|a_N z^N| &= \left| p(z) - \sum_{k=0}^{N-1} a_k z^k \right| \\
&\leq |p(z)| + \left| \sum_{k=0}^{N-1} a_k z^k \right| \\
&\leq |p(z)| + \frac{1}{2} |a_N| L^N,
\end{aligned}
$$

where the first inequality is the triangle inequality and the second is because $|z| > L$. Choose L_1 such that

$$\frac{1}{2} |a_N| L_1^N > |a_0|.$$

Let $L = \max(L_0, L_1)$, and let S be the corresponding closed square. The function $|p|$ is continuous on S, so it attains its minimum at some point, α_1, say, by Theorem 9.32. On the boundary of S, we know

$$|p(z)| \geq \frac{1}{2} |a_N| L^N > |a_0| = |p(0)|.$$

Therefore α_1 must be in the interior of S. By D'Alembert's lemma, we must have $p(\alpha_1) = 0$, or else there would be a nearby point ζ, also in S, where $|p(\zeta)|$ was smaller than $|p(\alpha_1)|$. So α_1 is a root of p.

(ii) Now we apply Lemma 9.35 to conclude that we can factor p as

$$p(z) = (z - \alpha_1)q(z),$$

where q is a polynomial of degree $N-1$. By a straightforward induction argument, we can factor p into N linear factors.

(iii) Uniqueness is obvious. The number c is the coefficient a_N. The numbers a_k are precisely the points at which the function p vanishes, as it follows from Proposition 9.16 that the product of finitely many complex numbers can be 0 if and only if one of the numbers is itself 0. $\square$

9.5 Application to Real Polynomials

If p is a polynomial in $\mathbb{R}[x]$, it follows from the Fundamental Theorem of Algebra that it does have roots, but they may be complex. If it has complex roots, they must occur in complex conjugate pairs.

THEOREM 9.42. *Let $p \in \mathbb{R}[x]$. Let α be a root of p. Then so is $\bar{\alpha}$.*

PROOF. Let $p(x) = \sum_{k=0}^{N} a_k x^k$. Then

$$p(\alpha) = \sum_{k=0}^{N} a_k \alpha^k = 0,$$

so

$$p(\bar{\alpha}) = \sum_{k=0}^{N} a_k \bar{\alpha}^k = \overline{p(\alpha)} = 0.$$

$\square$

Let $\alpha = a + ib$. Then

$$
\begin{aligned}
(x - \alpha)(x - \bar{\alpha}) &= (x - (a + ib))(x - (a - ib)) \\
&= x^2 - 2ax + a^2 + b^2 \\
&= (x - a)^2 + b^2.
\end{aligned}
\tag{9.43}
$$

So applying the Fundamental Theorem of Algebra to the real polynomial p, we first factor out the real roots, and for each pair of complex conjugate roots we get a factor as in (9.43). Thus we get the following theorem.

THEOREM 9.44. *Let* $p \in \mathbb{R}[x]$ *be a polynomial of degree* N. *Then* p *can be factored into a product of linear factors* $(x - c_k)$ *and quadratic factors* $((x - a_k)^2 + b_k^2)$:

$$
p(x) = c \left(\prod_{k=1}^{N_1} (x - c_k) \right) \left(\prod_{j=1}^{N_2} ((x - a_j)^2 + b_j^2) \right)
$$

for some (not necessarily distinct) real numbers c, c_j, a_j, b_j. *We have* $N_1 + 2N_2 = N$, *and the factoring is unique, up to ordering and replacing any* b_j *by* $-b_j$.

9.6 Further Remarks

In Chapter 5 we defined cosine and sine in terms of power series. In Section 9.2, we interpreted them geometrically and used trigonometric identities. Showing that the power series and the trigonometric interpretation are really describing the same function is part of a course in complex analysis.

There are two main ingredients to a first course in complex analysis. The first is to show that if a function f has a derivative everywhere on some open disk, in the sense that

$$
\lim_{z \to z_0} \frac{f(z_0) - f(z)}{z_0 - z}
$$

exists, then the function is automatically analytic, that is, expressible by a convergent power series. This is not true for real functions and explains much of the special nature of complex differentiable functions.

The second part of the course concerns evaluating contour integrals of complex differentiable functions. This is useful not only in its own right, but in applications to real analysis, such as inverting the Laplace transform or evaluating definite integrals.

A good introduction to complex analysis is the book by Donald Sarason [**8**].

9.7 Exercises

EXERCISE 9.1. What are the primitive fourth roots of unity?

EXERCISE 9.2. Show that if ω is any n^{th} root of unity other than 1, then $1 + \omega + \omega^2 + \cdots + \omega^{n-1} = 0$.

EXERCISE 9.3. How many primitive cube roots of unity are there? How many primitive sixth roots? How many primitive n^{th} roots for a general n?

EXERCISE 9.4. Redo Example 9.8 to get all three roots from the Tartaglia-Cardano formula.

EXERCISE 9.5. Let $p(x) = x^3 + 3x + \sqrt{2}$. Show without using the Cardano-Tartaglia formula that p has exactly one real root. Find it. What are the complex roots?

EXERCISE 9.6. Fill in the proof of Proposition 9.31.

EXERCISE 9.7. Let $g : G \to \mathbb{C}$ be a continuous function on $G \subseteq \mathbb{C}$. Show that $\Re(g)$, $\Im(g)$ and $|g|$ are continuous. Conversely, show that the continuity of $\Re(g)$ and $\Im(g)$ imply the continuity of g.

EXERCISE 9.8. Show that every continuous real-valued function on a closed, bounded subset of $\mathbb{C}$ attains its extrema.

The Greek Alphabet

Lower-case	Upper-case	Name
α	A	alpha
β	B	beta
γ	Γ	gamma
δ	Δ	delta
ε	E	epsilon
ζ	Z	zeta
η	H	eta
θ	Θ	theta
ι	I	iota
κ	K	kappa
λ	Λ	lambda
μ	M	mu
ν	N	nu
ξ	Ξ	xi
o	O	omicron
π	Π	pi
ρ	P	rho
σ	Σ	sigma
τ	T	tau
υ	Υ	upsilon
ϕ	Φ	phi
χ	X	chi
ψ	Ψ	psi
ω	Ω	omega

APPENDIX B

Axioms of Zermelo-Fraenkel
with the Axiom of Choice

Russell's paradox (Section 1.7) demonstrates that the General Comprehension Principle is false, as it gives rise to a contradiction. So how do we decide whether a definable collection is a set? This question engendered a program to *axiomatize* set theory with the objective of producing uniform assumptions about sets that satisfied numerous constraints:

- The axioms are understandable and intuitively sound. We must be able to recognize when a statement about sets is an axiom.
- The axioms are sufficient to derive the standard theorems of mathematics.
- The axioms are not redundant. That is, no axiom can be derived from the remaining axioms.
- Every mathematical statement about sets is either provable or refutable from the axioms.
- The axioms are logically consistent and hence do not give rise to a contradiction.

As we shall discuss later, no collection of axioms can simultaneously achieve these objectives. First we give the axioms on which mathematicians ultimately settled, the Axioms of Zermelo-Fraenkel (with the Axiom of Choice):

(1) **Extensionality.** If sets X and Y have the same elements, then $X = Y$.

(2) **Pairing.** For any sets X and Y, there is a set $Z = \{X, Y\}$.

(3) **Union.** Let X be a set of sets. Then there is a set

$$\{x \mid (\exists Y \in X)\, x \in Y\}.$$

(4) **Power Set.** If X is a set, then the collection of all subsets of X is a set.

(5) **Infinity.** There is an inductive set.

(6) **Schema of Separation.** If $P(x, y_1, \ldots, y_n)$ is a formula with $n + 1$ variables, and $X, X_1, \ldots, X_n$ are sets, then there is a set

$$\{x \in X \mid P(x, X_1, \ldots, X_n)\}.$$

(7) **Schema of Replacement.** If F is a function on arbitrary collections, X is a set, and $f = F|_X$, then the range of f is a set.

(8) **Regularity.** Let X be a set. Then there is no infinite sequence of elements of X, $\langle x_i \rangle$, such that for all $n \in \mathbb{N}$, $x_{n+1} \in x_n$.

(9) **Choice.** Let X be a set of nonempty sets. Then there is a function f with domain X such that for all $x \in X$, $f(x) \in x$.

The axioms of Zermelo-Fraenkel with the Axiom of Choice are referred to as ZFC. (The Axiom of Choice is referred to as AC, and the Zermelo-Fraenkel axioms without the Axiom of Choice are referred to as ZF.) There are seven axioms and two axiom *schemata*. The schemata give infinitely many axioms. The Extensionality Axiom characterizes set identity. It says that a set is defined by its members. The Axioms of Pairing, Union, and Power Set guarantee that collections built from sets with these set operations will be sets. The Axiom of Infinity implies that the natural numbers are a set. The Schema of Separation says that any subset of a given set defined by a formula is a set. It is a weakened version of the General Comprehension Principle. The

Schema of Replacement says that given a function, F, on arbitrary collections (not necessarily sets) and a set X, the range of $F|_X$ is a set. The Axiom of Regularity is a technical axiom that implies that no set may be a member of itself.

The Axiom of Choice is different from the other axioms in that it does not claim that a definable object in the universe of sets is also a set. Rather, it implies the existence of a function without specifying the function. If X is a set and f is the function with domain X whose existence is guaranteed by the Axiom of Choice, then f is called a choice function for X. The Axiom of Choice is logically equivalent to axioms that are frequently used in arguments in many branches of mathematics. For instance, the Axiom of Choice is equivalent to the claim that every set may be well-ordered (the Well-Ordering Principle). The axiom gave rise to interesting paradoxes that caused some mathematicians to question its validity. It was proved by Kurt Gödel that if the axioms of Zermelo-Fraenkel without Choice were logically consistent, then the axioms of Zermelo-Fraenkel with Choice were logically consistent. There were a few occasions in Chapter 6 when we invoked the Axiom of Choice. There were occasions (e.g., Cantor's Theorem) in which the axiom was actually necessary but discussing it would have been unacceptably confusing. The axiom is considered necessary by most mathematicians. For instance without it, or some logically equivalent axiom, we cannot even conclude that any pair of sets can be compared (i.e., for any sets X and Y, either $X \preceq Y$ or $Y \preceq X$).

Does ZFC achieve the objectives of an axiomatization of set theory? The axioms are generally intuitive with the possible exceptions of AC and the Regularity Axiom. It is also known that if ZFC without the Regularity Axiom is logically consistent, then ZFC with the Regularity

Axiom is logically consistent. Mathematicians assume ZFC almost universally without giving it too much consideration. The axioms of ZFC have been sufficient for proving the theorems of standard mathematics.

We say that a set is *decidable* (or *recursive*) if membership in the set can be determined by rote computation. For instance, the set of even integers is decidable—you can use the division algorithm to check whether an integer is divisible by 2. ZFC is a recursive set of axioms. Indeed it is necessary that a set of axioms be recursive to be of any practical use. According to Gödel's First Incompleteness theorem, any decidable set of axioms in which one can do arithmetic will be logically incomplete. That is, there are statements in the language of the axioms that are neither provable nor refutable from the axioms. It is not known, nor can it be known by a mathematical proof (using ZFC), whether ZFC is logically consistent. The consistency of a decidable set of axioms in which one can do arithmetic cannot be a logical consequence of those axioms. This result is known as Gödel's Second Incompleteness theorem and is one of the great results of twentieth-century mathematics.

For a good treatment of set theory at an undergraduate level, see Y. Moschovakis's book [**5**].

Bibliography

[1] M. Aigner and G. M. Ziegler. *Proofs from the Book* (Berlin: Springer, 2003).

[2] J. B. Fraleigh. *A First Course in Abstract Algebra* (Reading, MA: Addison-Wesley, 1982).

[3] I. N. Herstein. *Abstract Algebra* (New York: Wiley, 1999).

[4] I. Lakatos. *Proofs and Refutations: The Logic of Mathematical Discovery* (Cambridge, UK: Cambridge University Press, 1976).

[5] Y. N. Moschovakis. *Notes on Set Theory* (Berlin: Springer, 1994).

[6] The MacTutor History of Mathematics Archives, *http://www-groups.dcs.st-and.ac.uk/~history/index.html*

[7] W. Rudin. *Principles of Mathematical Analysis* (New York: McGraw-Hill, 1976).

[8] D. Sarason. *Notes on Complex Function Theory* (New Delhi: Hindustan Book Agency, 1998).

[9] E. M. Stein and R. Shakarchi. *Fourier Analysis* (Princeton, NJ: Princeton University Press, 2003).

Index

Page numbers appear in bold for pages where the term is defined.